Bioactive Components in Fermented Foods and Food By-Products

Special Issue Editors

Vito Verardo

Ana María Gomez-Caravaca

Giulia Tabanelli

MDPI • Basel • Beijing • Wuhan • Barcelona • Belgrade • Manchester • Tokyo • Cluj • Tianjin

Special Issue Editors
Vito Verardo
University of Granada
Spain

Ana María Gomez-Caravaca
University of Granada
Spain

Giulia Tabanelli
University of Bologna
Italy

Editorial Office
MDPI
St. Alban-Anlage 66
4052 Basel, Switzerland

This is a reprint of articles from the Special Issue published online in the open access journal *Foods* (ISSN 2304-8158) (available at: https://www.mdpi.com/journal/foods/special_issues/fermented_foods_bioactive_compounds).

For citation purposes, cite each article independently as indicated on the article page online and as indicated below:

LastName, A.A.; LastName, B.B.; LastName, C.C. Article Title. *Journal Name* **Year**, *Article Number*, Page Range.

ISBN 978-3-03928-851-9 (Pbk)
ISBN 978-3-03928-852-6 (PDF)

Contents

About the Special Issue Editors

Vito Verardo is a food technologist (University of Bologna, Italy, 2004) with a PhD in Food Science (University of Bologna, Italy, 2009). He is currently Assistant Professor at the department of Nutrition and Food Science, and Researcher at the Institute of Nutrition and Food Technology 'José Mataix' at University of Granada. His main area of expertise is in the fields of food chemistry and food technology. He has acquired multidisciplinary knowledge on the characterization of bioactive and potentially toxic compounds in food and biological samples using advanced analytical techniques; the development of new food enriched in bioactive compounds and the characterization and valorization of food by-products; the study of lipid oxidation and strategies to limit it; and the use of green technologies for nutraceutical production. His level of research activity is reflected in his participation in 21 projects (4 financed by the European Community (two as PI and responsible for the WP)) financed by competitive calls (where he is PI for 5), and 12 contract projects financed by companies (being PI in one of them). He has published more than 80 articles in scientific journals and 11 book chapters in international editorials and 2 in national editorials.

Ana María Gomez-Caravaca is Associate Professor at the Analytical Chemistry Department of the University of Granada (Spain). During her career, her research has focused on the characterization of bioactive compounds (phenolic compounds, sterols, tocopherols, fatty acids, terpenes, alkylphenols, etc.) in plant matrices and in biological fluids, the development of foods enriched with bioactive compounds, the use of green technologies, and the study and characterization of food by-products for their use as ingredients in the food and pharmaceutical industries. The studied matrices include olive oil, olives and their by-products, tropical fruits and their by-products, and cereals. The research work has been carried out through the use of advanced analytical techniques such as gas chromatography coupled to FID and mass detectors, liquid chromatography, liquid nanochromatography, and capillary electrophoresis coupled to detectors such as DAD, fluorimeter, ion trap mass spectrometry, time of flight mass spectrometry, quadrupole time of flight mass spectrometry, and nuclear magnetic resonance (LC/nanoLC/CE–MS (TOF/Q-TOF/IT–NMR), among others. She has participated in 31 research projects funded in competitive calls (3 as PI). She has published 90 articles in internationally renowned journals, 8 chapters in international editorials, and 2 chapters in national editorials.

Giulia Tabanelli is Junior Assistant Professor at the Department of Agricultural and Food Sciences (DISTAL) of the University of Bologna from 2019, performing research in the field of food microbiology. She obtained her master's degree in Food Science and Technology, University of Bologna, in 2007, and PhD in Food Science and Biotechnology in 2011. Her research deals in particular with the role of microbiota in fermented foods, the physiology and metabolism of lactic acid bacteria, biogenic amine production, the selection of bacterial strains and evaluation of their performances for industrial use, the study of the antimicrobial characteristics of essential oils and their components, and physiological responses of the microorganisms subjected to sublethal stresses. Moreover, she studies the use of predictive microbiology to evaluate the microbial stability of foods in relation to applied control factors and the evaluation of the effects of fermentation processes on the quality and safety of fermented foods. She participated in numerous national and international research projects, including one financed by the European Community, financed in competitive calls, and she is involved in industrial research activities dedicated to the technological transfer to agri-food industry, being PI in some of them. She is the author and co-author of 56 publications in scientific journals indexed in the ISI Web of Science and SCOPUS databases (h index of 18 and a total of 968 citations as of 27 February 2020) and she is the co-author of 5 book chapters in international editorials.

foods

MDPI

Editorial

Bioactive Components in Fermented Foods and Food By-Products

Vito Verardo [1,2,*], Ana Maria Gómez-Caravaca [3,*] and Giulia Tabanelli [4,*]

1 Department of Nutrition and Food Science, University of Granada, Campus of Cartuja, 18071 Granada, Spain
2 Institute of Nutrition and Food Technology 'José Mataix', Biomedical Research Center, University of Granada, Avda del Conocimiento sn., 18100 Armilla, Granada, Spain
3 Department of Analytical Chemistry, University of Granada. Campus of Fuentenueva, Avda Fuentenueva s/n, 18071 Granada, Spain
4 Department of Agricultural and Food Sciences, University of Bologna, Viale Fanin 42, 40127 Bologna, Italy
* Correspondence: vitoverardo@ugr.es (V.V.); anagomez@ugr.es (A.M.G.-C.); giulia.tabanelli2@unibo.it (G.T.)

Received: 21 January 2020; Accepted: 23 January 2020; Published: 5 February 2020

Keywords: food fermentation; food by-products; bioactive compounds; lactic acid bacteria; lycopene; phenolic compounds; biogenic amines; γ-aminobutyric acid (GABA)

Food fermentation is one of the most ancient processes of food production that has historically been used to extend food shelf life and to enhance its organoleptic properties. However, some research has demonstrated that it can also increase the nutritional value and/or digestibility of food.

Firstly, microorganisms, and in particular Lactic Acid Bacteria (LAB), besides their role in acidification, are able to produce huge amounts of secondary metabolites with excellent health benefits and preservative properties (i.e., antimicrobial activity). Indeed, some microorganisms can increase the levels of several bioactive compounds (e.g., vitamins, antioxidant compounds, peptides, etc.). Secondly, fermented foods contain living organisms that contribute to the modulation of the host's physiological balance and gut microbiota, enriching, at the same time, the host's diet with new bioactive molecules.

Moreover, recent advances in fermentation are focused on food by-products; in fact, they are a source of potentially bioactive compounds that, after fermentation, could be used as ingredients for nutraceuticals and functional food formulations [1].

Because of this, understanding of benefits of food fermentation is a growing field of research in nutrition and food science.

This Special Issue aimed to present current knowledge and research trends concerning the use of fermentation technologies like the sustainable GRAS (Generally Recognized As Safe) process for food and nutraceutical production, to improve food quality.

In this context, Verni et al. [2] reviewed the effect of fermentation on the antioxidant compounds of cereals and legume-derived foods. They reported that the ability of fermentation to improve food antioxidant properties is strictly related to the metabolic activities of the starter used. Briefly, the fermentation processes are able to improve the bio-accessibility of phenolic compounds. Moreover, bioactive peptides resulted from bacterial and fungal proteolysis.

Several original papers focused on the use of food by-product fermentation in order to improve the extraction of bioactive compounds. Doan et al. [3] isolated a filamentous fungus (*Clerodendron cyrtophyllum*) from the root of *Clerodendron cyrtophyllum* Turcz; they grew this fungus on isoflavones rich soybean extract. Thanks to the high β-glucosidase production of *Clerodendron cyrtophyllum*, they were able to produce an isoflavones aglycones-rich soybean extract (e.g., genistein and daidzein), proposing this method for applications in the pharmaceutical and functional food industries. Another study carried out by Lordan and co-workers [4] assessed the antithrombotic activities of lipid extracts from brewing raw materials, by-products, wort, and beer. Briefly, they showed that the fermentation of a brewing industry's by-products could play a key role in increasing the anti-platelet-activating factor bioactivity of polar lipids. Simat and co-workers [5] also proposed aquaculture by-products as an alternative source of sustainable and profitable bioactive fish oils.

Tofalo et al. [6] studied the effect of traditional cheese fermentation on the accumulation of healthy (γ-aminobutyric acid or GABA) and toxic (biogenic amines (BA)) compounds. They confirmed a greater BA formation and proteolytic activity in cheese made by pig rennet than those made by calf and kid rennet. So, they proposed the selection of autochthonous amine-negative and amine-oxidizing LAB as a valuable strategy to decrease BA formation. However, high amounts of GABA were produced, and they were correlated with the use of ewe's milk, time of ripening, and type of coagulant.

Venturi and co-workers [7] selected two strains belonging to *Lactobacillus farciminis* and *Lactobacillus brevis* species and used them for amaranth bread production. Their results underlined the bread produced with these LAB showed higher antioxidant activity and total phenolic content compared to the control. Moreover, these strains were able to increase GABA concentration (up to 350%) in breads enriched with 20% amaranth flour.

Sevgili et al. [8] checked the use of different substrates to produce lycopene via *Blakeslea trispora* fermentation. They confirmed that the medium with natural oil showed more lycopene than the medium that contained only a carbon source. In fact, oils improved lycopene production and the highest lycopene concentration was obtained when the medium was added to sunflower and corn oils.

Finally, Fracassetti et al. [9] proposed a new analytical method for the simultaneous quantification of tryptophan (TRP), tryptophan ethylester (TEE), and melatonin isomers (MISs) in fermented foods, such as wine. A preconcentration of wine by Solid Phase Extraction (SPE) followed by high performance liquid chromatography (HPLC) analysis either with fluorescence or mass spectrometer detectors were applied. They suggested that this protocol could be a useful tool for monitoring the release of MEL and TEE, even when their concentration is very low (<0.02 μg/L), as well as to determine their presence and concentration during wine production and storage.

To conclude, the papers included in this Special Issue summarized the valuable use of fermentation technology to improve the content of bioactive compounds in foods and food by-products or to limit the presence of some toxic compounds, such as biogenic amines. At the same time, different analytical approaches were proposed in order to study the effect of fermentation on the food matrix and the possible metabolites that are produced. However, as suggested by several authors, more in vivo studies should be carried out to demonstrate the potential of fermentation to produce healthy foods.

Author Contributions: V.V., A.M.G.C. and G.T. wrote the paper. All authors have read and agreed to the published version of the manuscript.

Funding: This research received no external funding.

Acknowledgments: Vito Verardo thanks the Spanish Ministry of Economy and Competitiveness (MINECO) for "Ramon y Cajal" contract (RYC-2015-18795).

Conflicts of Interest: The authors declare no conflict of interest.

References

1. Martinez-Avila, G.C.G.; Aguilera, A.F.; Saucedo, S.; Rojas, R.; Rodriguez, R.; Aguilar, C.N. Fruit Wastes Fermentation for Phenolic Antioxidants Production and Their Application in Manufacture of Edible Coatings and Films. *Crit. Rev. Food Sci. Nutr.* **2014**, *54*, 303–311. [CrossRef] [PubMed]
2. Verni, M.; Verardo, V.; Rizzello, C.G. How Fermentation Affects the Antioxidant Properties of Cereals and Legumes. *Foods* **2019**, *8*, 362. [CrossRef] [PubMed]
3. Doan, D.T.; Luu, D.P.; Nguyen, T.D.; Hoang Thi, B.; Pham Thi, H.M.; Do, H.N.; Luu, V.H.; Pham, T.D.; Than, V.T.; Pham Thi, H.H.; et al. Isolation of Penicillium citrinum from Roots of Clerodendron cyrtophyllum and Application in Biosynthesis of Aglycone Isoflavones from Soybean Waste Fermentation. *Foods* **2019**, *8*, 554. [CrossRef] [PubMed]
4. Lordan, R.; O'Keeffe, E.; Tsoupras, A.; Zabetakis, I. Total, Neutral, and Polar Lipids of Brewing Ingredients, By-Products and Beer: Evaluation of Antithrombotic Activities. *Foods* **2019**, *8*, 171. [CrossRef] [PubMed]

5. Simat, V.; Vlahović, J.; Soldo, B.; Skroza, D.; Ljubenkov, I.; Generalić Mekinić, I. Production and Refinement of Omega-3 Rich Oils from Processing By-Products of Farmed Fish Species. *Foods* **2019**, *8*, 125. [CrossRef] [PubMed]
6. Tofalo, R.; Perpetuini, G.; Battistelli, N.; Pepe, A.; Ianni, A.; Martino, G.; Suzzi, G. Accumulation γ-Aminobutyric Acid and Biogenic Amines in a Traditional Raw Milk Ewe's Cheese. *Foods* **2019**, *8*, 401. [CrossRef] [PubMed]
7. Venturi, M.; Galli, V.; Pini, N.; Guerrini, S.; Granchi, L. Use of Selected Lactobacilli to Increase γ-Aminobutyric Acid (GABA) Content in Sourdough Bread Enriched with Amaranth Flour. *Foods* **2019**, *8*, 218. [CrossRef] [PubMed]
8. Sevgili, A.; Erkmen, O. Improved Lycopene Production from Different Substrates by Mated Fermentation of Blakeslea Trispora. *Foods* **2019**, *8*, 120. [CrossRef] [PubMed]
9. Fracassetti, D.; Vigentini, I.; Lo Faro, A.F.F.; De Nisi, P.; Foschino, R.; Tirelli, A.; Orioli, M.; Iriti, M. Assessment of Tryptophan, Tryptophan Ethylester, and Melatonin Derivatives in Red Wine by SPE-HPLC-FL and SPE-HPLC-MS Methods. *Foods* **2019**, *8*, 99. [CrossRef] [PubMed]

foods

MDPI

Review

How Fermentation Affects the Antioxidant Properties of Cereals and Legumes

Michela Verni [1,*], Vito Verardo [2,3] and Carlo Giuseppe Rizzello [1]

1 Department of Soil, Plant and Food Science, University of Bari Aldo Moro, 70126 Bari, Italy
2 Department of Nutrition and Food Science, University of Granada, Campus Universitario de Cartuja, E-18071 Granada, Spain
3 Institute of Nutrition and Food Technology 'José Mataix', Biomedical Research Centre, University of Granada, Avenida del Conocimiento s/n, E-18071 Granada, Spain
* Correspondence: michela.verni@uniba.it; Tel.: +39-080-544-2950

Received: 24 July 2019; Accepted: 22 August 2019; Published: 24 August 2019

Abstract: The major role of antioxidant compounds in preserving food shelf life, as well as providing health promoting benefits, combined with the increasing concern towards synthetic antioxidants, has led the scientific community to focus on natural antioxidants present in food matrices or resulting from microbial metabolism during fermentation. This review aims at providing a comprehensive overview of the effect of fermentation on the antioxidant compounds of vegetables, with emphasis on cereals- and legumes- derived foods. Polyphenols are the main natural antioxidants in food. However, they are often bound to cell wall, glycosylated, or in polymeric forms, which affect their bioaccessibility, yet several metabolic activities are involved in their release or conversion in more active forms. In some cases, the antioxidant properties in vitro, were also confirmed during in vivo studies. Similarly, bioactive peptides resulted from bacterial and fungal proteolysis, were also found to have ex vivo protective effect against oxidation. Fermentation also influenced the bioaccessibility of other compounds, such as vitamins and exopolysaccharides, enabling a further improvement of antioxidant activity in vitro and in vivo. The ability of fermentation to improve food antioxidant properties strictly relies on the metabolic activities of the starter used, and to further demonstrate its potential, more in vivo studies should be carried out.

Keywords: Lactic acid bacteria; fungi; phenolic compounds; bioactive peptides; grains

1. Antioxidant Compounds in Food Matrices

The increasing interest towards healthier food and lifestyles has steered the scientific community to pay great attention to the field of free radicals and antioxidant compounds. Free radicals are atoms, molecules, or ions containing an unpaired electron, which makes them unstable and highly reactive [1]. The generation of free radicals in our body occurs as consequence of exposure to different physiochemical conditions or pathological states. Free radicals are responsible for damaging biologically relevant molecules and lipids, nucleic acids, and proteins are the major targets [2]. They are generated continuously; however, they are also involved in inter-cellular and intra-cellular signaling systems and enzymatic reactions essential to intermediary metabolic processes of life [3], therefore, their daily production must be balanced. If free radicals overcome the body's ability to regulate them, and the ratio between free radical generation and antioxidant defenses is unbalanced, a condition known as oxidative stress ensues. Being the oxidative stress responsible for an enormous number of conditions, including cancer, cardiovascular and neurodegenerative diseases, atherosclerosis, inflammatory state, and many others [2,4], preventing it by enhancing the intake of dietary antioxidants represents the most feasible way of protection against free radicals. Antioxidants need to be capable of delaying or inhibiting the oxidation of a substrate, yet an important property they should also have, is the ability to form a new radical that is stable through intramolecular hydrogen bonding on further oxidation [1]. Several mechanisms of action can be responsible for their activity. Antioxidants can (i) scavenge species that initiate peroxidation, (ii) donate a hydrogen or an electron, (iii) chelate metal ions preventing the generation of reactive species or lipid peroxides de-composition, (iv) quench the radical O_2^- preventing peroxides formation, (v) breaking autoxidative chain reaction, (vi) inhibit pro-oxidative enzymes, and/or (vii) reduce localized O_2 concentrations [1,5]. From a technological point of view, antioxidants are designed to prevent food from spoiling through oxidation, thus reducing loss of nutrients, and maintaining texture, color pigments, taste, freshness, functionality, and aroma [3]. Therefore, antioxidants are an important category of food preservatives and can be divided into natural or synthetic. Natural antioxidants include flavonoids, phenolic acids, carotenoids, and tocopherols [1]; however, other protein derived compounds, such as amino acids and bioactive peptides, have received great attention for their displayed antioxidant properties [6–8]. Both natural and synthetic antioxidants, act by similar mechanisms and their chemical structure and polarity influence the antioxidant activity [5]. Antioxidants efficiency generally increases with the increase of their concentration; however, the dependence is not linear, and when the maximum activity is reached, it may also decrease [9]. Except for carotenes, tocopherols, and their esters, natural antioxidants are mostly much more polar than synthetic ones. They are also less active and substrate specific, and their antioxidant activity depends highly on synergists factors [9].

This review aims at providing a comprehensive overview of the effect of fermentation on the antioxidant activity of vegetable matrices. The literature is full of articles claiming the potential of food fermentation in improving antioxidant properties; however, the method used to quantify phenolic compounds, unless performed by chromatographic analysis is the classic Folin-Ciocalteu, which suffers from several drawbacks. The test is sensitive to pH, temperature, and reaction time; inorganic and non-phenolic organic substances, including reducing agents, react with Folin reagent, causing overestimations of the phenolic content [10,11]. For this reason, all the examples reported in this review were chosen among those in which a characterization of the phenolic profile was performed. Moreover, among vegetable matrices, only grains were considered since (i) diets in developing countries are primarily based on cereals and legumes, whereas in Western society there is an increasing interest in strictly vegetarian diets [12]; (ii) cereals are often consumed as refined products, yet most of the bioactive compounds are in the outer layers of the grains, and get lost during milling [13,14]; (iii) legumes consumption is often limited by the presence of antinutritional compounds which fermentation has proven to diminish [13,14]. After an extensive research on multiple databases, 76 research articles, half of which published in the last five years, highlighting the effect of fermentation on grains antioxidant properties were selected for this review. Respectively, 39 and 17 papers reported phenolic and proteic compounds as responsible for the antioxidant activity. Moreover, such activity was demonstrated in vivo in about 15% of the studies considered.

1.1. Phenolic Compounds

The antioxidant activity of phenolic compounds lies in their ideal chemical structure, facilitating the hydrogen or electron donation from hydroxyl groups positioned along the aromatic ring and conferring radical scavenging activities and metal-chelating potential. Phenolics have the ability of stabilizing and delocalizing the un-paired electron within their aromatic rings [15]. Phenolics are composed of at least one aromatic ring with at least one hydroxyl group and may be classified based on the number of phenol rings and the structural elements that are bound to the rings [16]. Phenolic acids, flavonoids, tannins, stilbenes, and lignans are the main groups of phenolics.

Phenolic acids are divided into hydroxybenzoic acids and hydroxycinnamic acids and usually act as antioxidants by trapping free radicals. Flavonoids, instead, can scavenge free radicals and chelate metals [5,15]. The common characteristic of flavonoids is the basic 15-carbon flavan structure. They are arranged in three rings (A, B, and C) and the different classes vary for the level of saturation of the C ring, whereas compounds within the same class differ for the level of substitution of A and B rings [5]. Polyphenols stability and free radical scavenging potential depend on both the number and location of the free OH group. The antioxidant activity increases with the increase of the hydroxyl groups, especially if positioned in ortho-3,4 [17].

Based on their chemical structure, tannins are defined as hydrolysable or condensed (proanthocyanidins). Condensed tannins are oligomers and polymers of flavan-3-ols, whereas hydrolysable tannins are glycosylated gallic and ellagic acid derivatives [18]. It was proven that free radicals scavenging activity of both hydrolysable and condensed tannins involves a fast and a slow step [19]. The fast scavenging reaction is inhibited by complexation of the tannin with protein, a very tight bond between the phenolic group of tannins and the NH group of proteins, which prevents their hydrolysis and digestion in the stomach [19]. Nevertheless, the overall capacity of the tannin-protein complex for scavenging seems to be similar to that of the free tannin [18,19].

1.2. Antioxidant Peptides and Protein Derivatives

Besides their nutritional, physicochemical, and sensory properties, proteins can be responsible for health promoting benefits, mainly attributed to biologically active peptides [20]. Bioactive peptides are produced by digestive enzymes during gastrointestinal digestion, or by proteolytic enzymes during food processing (ripening, fermentation, cooking), storage, or in vitro hydrolysis [7]. These peptides may play various roles (antimicrobial, antihypertensive, hypocholesterolemic, immunomodulatory, antioxidative, antithrombotic, antitumoral) and can be released from native proteins that derive from vegetable or animal matrices [7,8].

Peptides displaying antioxidant activity usually contain 5–16 amino acid residues [21]. The exact mechanism behind it has not fully been understood, yet several studies reported that they are inhibitors of lipid peroxidation, scavengers of free radicals and chelators of transition metal ions. [8,21]. Tyrosine, tryptophan, methionine, lysine, cysteine, and histidine are examples of amino acids displaying antioxidant activity. Synthesized peptides containing the active fragments have been proven to inhibit lipid peroxidation, while the tripeptides, Tyr-HisTyr, and Pro-His-His were found to be effective in stabilizing radical and non-radical oxygen species, including peroxynitrite and lipid peroxide [22]. Amino acids with aromatic residues can donate protons to electron deficient radicals, whereas histidine-containing peptides, thanks to the imidazole group, have been found to have hydrogen-donating, lipid peroxyl radical trapping, and metal ion-chelating abilities [8]. On the other hand, sulphur containing amino acids, have antioxidant action due to the direct interaction of thiol group with radicals. Cysteine and homocysteine inhibit LDL (low-density lipoprotein) oxidation by hemin and copper, methionine residues, instead, scavenge oxidizing agents [6]. Besides the presence of the proper amino acids, their specific positioning in the sequence plays an important role in antioxidant activity of peptides as well as other factors, such as the structure, amino acids configuration, hydrophobicity, and concentration [8].

1.3. Synthetic Antioxidants

Synthetic antioxidants were developed to prolong food shelf life but also because of the need to have a standard measurement system to compare with natural antioxidants. There are numerous compounds used in food, animal, and cosmetic applications to prevent oxidation; some also have antifungal properties and possess at least one phenolic ring in the structure. Butylated hydroxyanisole (BHA), butylated hydroxytoluene (BHT), tert-butylhydroquinone, propyl gallate, octyl gallate, 2,4,5-Trihydroxy butyrophenone, nordihydroguaiaretic acid, and 4-Hexylresorcinol are some examples [1]. Among these, BHT and BHA, alone or in combination with others, are the most commonly employed [2] and synergistic effects were also reported [23]. Today, almost all processed foods contain synthetic antioxidants and, despite being reported safe in the past, several studies have addressed their danger for human health, among which liver, kidney, and lungs damages, mutagenesis, carcinogenesis, and many others [19]. For this reason, between 2011 and 2012, the European food safety authority re-evaluated their maximum levels of intake in adults and children [24,25]. Considering the increasing risk factors related to synthetic antioxidants, there has been a global trend toward the use and the search for effective natural substance as therapeutic antioxidants.

2. Bioaccessibility and Bioavailability of Antioxidant Compounds

The research around antioxidants has grown exponentially, but there are still limitations that need to be considered before the real potential of these molecules is properly appreciated. All the bioactive compounds, to exert their biological properties, must be available in the target tissue, which is why, when addressing a specific property, it is important to consider their absorption in the gastro-intestinal tract. The absorption depends on the compound bioaccessibility and bioavailability. Only polyphenols that reach the gut, released from the food matrix by the action of digestive enzymes and gut microbiota, are bioaccessible, and therefore, potentially bioavailable [26].

Among polyphenol classes, their physicochemical characteristics play a significant role in the overall availability. It was observed that the absorption of phenolic acids, having small-molecular weight, as well as flavones, catechins, and quercetin glucosides results easier compared to larger polyphenols such as proanthocyanidins, which need to be degraded into monomer or dimer units before being absorbed [27,28]. Anthocyanins can be absorbed as glycosides and appear as such in blood [29], whereas galloylated monomeric flavonols do not seem to undergo extensive metabolism [30]. Another important factor to consider when assessing polyphenols bioavailability is their metabolism and the biotransformation reactions they can undergo once absorbed. As a result of the changes in their structure they may or may not still exert the biological action [28].

As for peptides, their ability to resist enzymatic digestion in the gastrointestinal tract is necessary to ensure their bioactivity within the human body. Since peptides' potential properties strictly depend on the amino acid composition, the loss of amino acid residues as a result of proteolysis in the gastrointestinal tract can lead to changes in the activity. Although it is more likely that small molecules get absorbed without undergoing further digestion, studies have showed that peptides with higher molecular weight can pass to the plasma without modifications and it seems that the amino acids composition is a factor of key importance in the resistance during the digestion tract [31].

Even though the use of in vitro tests has generated a controversy over the last years, these methods are still of great importance in the selection of potential antioxidant compounds. Therefore, in vitro assays are still necessary to screen among the thesis, yet major effort should be put into validate in vivo the bioaccessibility of such compounds. Despite the high reliability recognized to the in vivo studies, the problems related to the approval for the ethical committee, the long times and the high costs and the dependence on the individual response, pushed the development of in vitro methods simulating human digestion. The most widely used procedure for screening bioaccessibility is the in vitro static gastro-intestinal method. However, this method does not include a colonic phase where compounds may be metabolized by the colonic microbiota. A more reliable assessment can be obtained with dynamic gastro-intestinal models, which include the biological environment of the intestine [26].

3. Estimation of the Antioxidant Activity in Foods

3.1. *In Vitro Assays*

Although dozens of methods exist to investigate antioxidant activity, only a few are commonly employed to assess food antioxidant potential (for a review see Alam et al. [4]). Among scavenging activity assays, ABTS (2,20-azino-bis (3-ethylbenzothiazoline-6-sulfonic acid)) and DPPH (2,2-diphenyl-1-picrylhydrazyl) are colorimetric assays where the radicals decolorize in the presence of antioxidants [1]. In the hydroxyl radical scavenging, activity fluorescein is used as a probe and the fluorescence decay curve is monitored in the presence and absence of the antioxidants. ORAC (oxygen radical absorbance capacity) is a test performed using Trolox (a water-soluble analog of Vitamin E) as a standard and is based on the generation of free radical and measurement of decrease in fluorescence in the presence of free radical scavengers [4]. Another assay commonly used is the ferric reducing-antioxidant power

(FRAP), characterized by the reduction of Fe^{3+} to Fe^{2+} followed by the alteration of color from yellow to blue and analyzed through a spectrophotometer [4]. Several other assays, among which thiobarbituric reactive substances (TBARS) and glutathione peroxidase (GSHPx) methods, evaluate the inhibition of lipid peroxidation [1]. Considering the limitations of in vitro methods: (i) presence in the extracts of pigments and fluorophores interfering with absorbance and fluorescence readings, (ii) failure to evaluate radical scavenging rate, and (iii) lack of biological relevance due to the use of artificial radicals not found in food or biological systems; antioxidant activity should not be concluded based on a single antioxidant test model [11].

3.2. Ex Vivo Assays

Since in vitro assays fail to predict the antioxidant activity in vivo and testing a substance directly on animals or human is not an easy approach, methods comprising cellular models for a rapid initial screening have been developed. The hemolysis inhibition assay includes the use of plasma as substrate of oxidation. When exposed to ROS (reactive oxygen species), the oxidation of protein (hemoglobin) and lipid (mainly cholesteryl ester) begins, leading to destruction of the cell shape and membrane structure and ultimately hemolysis. The degree of hemolysis is determined spectrophotometrically measuring the concentration of released hemoglobin in the solution and the inhibition of hemolysis by antioxidants calculated by comparing with a control containing no antioxidants [11]. In cell culture models, human or animal cell lines such as keratinocyte or fibroblasts are subjected to oxidative-induced stress and then incubated for a certain amount of time with the substance to test. At the end of incubation, MTT (3-(4,5-dimethylthiazol-2-yl)-2,5- diphenyltetrazolium bromide) assay, which evaluates the ability of succinate dehydrogenase to convert MTT into formazan crystals in viable cells is performed [32,33]. Otherwise, in cellular antioxidant assays, a fluorescent probe is introduced into the cell cultures and in the presence of ROS or RNS (reactive nitrogen species), the substance is excited emitting fluorescence. The fluorescence intensity measured is proportional to the level of oxidation. Antioxidants absorbed into the cells scavenge the radicals, resulting in lower degree of oxidation observed as attenuated fluorescence increase [11]. Besides cellular assays, for the ex vivo evaluation of antioxidant activity, other biological systems exist, such as the inhibition of LDL and DNA oxidation assays. In the LDL-cholesterol assay, oxidation is induced by transition metals or peroxyl radical, and LDL is incubated with the samples. Then, the extent of the oxidation is determined by measuring the generated amount of lipid peroxides and by the TBARS assay [4]. DNA oxidation assay is based on a similar principle, but in this case, the DNA strand breaking is induced by hydroxyl or peroxyl radicals because they are the major sources responsible for DNA oxidative damage, especially mitochondrial [11].

3.3. In Vivo Assays

In vivo protocols commonly include the administration of antioxidants to testing animals for a specified period of time, after which the animals are sacrificed, and blood or tissues analyzed [4]. The lipid peroxidation (LPO) assay, which measures spectrophotometrically the end products of LPO process in the tested tissue, is one of the most used. GSHPx, instead, catalyzes the reduction of hydroperoxides. GSHPx measurement is performed especially in patients who are under oxidative stress; low activity of this enzyme is one of the early consequences of a disturbance of the prooxidant/antioxidant balance. While the FRAP assay, which was originally applied to plasma, is one of the most rapid tests and very useful for routine analysis [4].

4. Effect of Microbial Fermentation on the Antioxidant Activity

4.1. Metabolic Activities Affecting Phenolics

Plant phenolics are known to possess antimicrobial properties against bacteria, fungi, and yeasts; therefore, the ability to metabolize them comes from the need of detoxifying such compounds that, if present at high concentrations negatively affect the integrity of the cell wall and membrane, dissipate the pH gradient, delay the metabolism of carbohydrates and denature proteins [34–37]. Whether it is carried out by fungi, yeasts, or bacteria, microbial fermentation has an impact on the phenolic compounds characterizing food matrices and metabolic activities are species- or strains-specific and depend on their portfolio of enzymes. A schematization of the major effects of fermentation on phenolic compounds is reported in Table 1.

Table 1. Main effect of polyphenols metabolism on the antioxidant activity of fermented cereals and legumes.

Matrix	Microorganisms Employed	Process Parameters	Effect	Reference
Wheat bran	Baker's yeast	20 °C for 20 h	Release of phenolic acids and improved bioaccessibility and colonic metabolism of phenolic acids	[38]
	Spontaneous fermentation conducted mainly by *Lactobacillus*, *Leuconostoc* and *Pediococcus* spp.	Backslopping for 13 days at 18 °C	Release of ferulic acid	[39]
	Baker's yeast and LAB	20 °C for 24 h in anaerobic condition	Conversion of ferulic and caffeic acids into their derivatives and increase in sinapic acid.	[40]
	Hericium erinaceus and enzymes	25 °C for 7 days	Release of ferulic acid	[41]
Kamut bread	Baker's yeast and spontaneously fermented sourdough	30 °C for 1.5 h	Response to oxidative stress in in vivo studies with rats.	[42]
Rice bran	*Rhizopus oryzae* CTT 1217	30 °C for 120 h	Increase ferulic acid and DPPH scavenging activity. Inhibition of peroxidase and polyphenol oxidase.	[43]
	Rhizopus oligosporus F0020 *Monascus purpureus* F0061	32 °C for 12 days	Release of phenolic acids and increase of FRAP	[44]
Rye bran	*Bacillus subtilis* subsp. *subtilis* NRRL NRS-744	37 °C for 8 days	Release of phenolic acids	[45]
	Baker's yeast	25 °C for 14 h	Increase of phenolic acids	[46]
	Baker's yeast	20 °C for 24 h	Release of phenolic acids but no improved bioaccessibility in in vitro digestive systems	[47,48]
Sorghum flour	*Lactobacillus plantarum* FUA3171, *Lactobacillus casei* FUA3166, *Lactobacillus fermentum* FUA3165 *Lactobacillus reuteri* FUA3168	34 °C for 24 h	Release of phenolic acids and flavonoids	[49]
Tef pancake	Spontaneously fermented sourdough	25 °C for 24-120 h	Solubilization of bound phenolics and improved antioxidant potentials on FRAP and ABTS	[50]
Malt based beverage	*Lb. plantarum* Lp758, Lp765, Lp725, *Lactobacillus brevis* Lb986, *Lactobacillus amylolyticus* LaTL3, LaT15	30 °C for 72 h	Decarboxylation of phenolic acids.	[51]

Table 1. *Cont.*

Matrix	Microorganisms Employed	Process Parameters	Effect	Reference
Soy	*Bacillus subtilis* SHZ	30 °C for 36 h	Superoxide radical scavenging activity and reducing power potential	[52]
	B. subtilis MTCC5480, MTCC1747	42 °C for 24 h	Increase in the antioxidant activity due to both phenolic compounds and peptides	[53]
	Aspergillus oryzae ATCC 1011 *Rhizopus oryzae* ATCC 24563 *Bacillus subtilis* ATCC 6051	30 °C for 48 h	Increase in phenolic acids and flavonoids	[54]
	Bacillus pumilus HY1	37 °C for 60 h	Increase in phenolic acids, flavonoids and tannins monomeric forms	[55]
	Aspergillus awamori *Aspergillus niger* *Aspergillus niveus*	30 °C for 5 days	Increase in phenolic acids and flavonoids as consequence of β-glucosidase activity	[56]
	R. oligosporus	Room temperature for 20 days	Increase in phenolic acids and flavonoids	[57]
	Lentinus edodes CY-35	Room temperature for 50 days	Increase in phenolic acids and flavonoids as consequence of laccase and β-glucosidase activities	[58]
	Bacillus natto	37 °C for 48 h	Increase in phenolic acids and flavonoids	[59]
	Lactobacillus plantarum CECT 748 T	30 °C for 48 h	Increase in phenolic acids and flavonoids	[60]
	Pool of selected LAB	37 °C for 48 h	Increase in phenolic acids, flavonoids, saponins, phytosterols, and tocopherols	[61]
	Lactobacillus casei *Lactobacillus acidophilus*	32 °C for 15 h	Increased of the aglycones/glycosylated isoflavones ratio and bound phenolics	[62]
	Lactobacillus casei 0979 after germination	30 °C for 24 h	Increase in isoflavones glycosides and aglycones	[63]
	Kefir grains containing LAB and yeasts	30 °C for 24 h	Increase of isoflavones and improved antioxidant activities on DPPH and ABTS	[64]
	Lb. plantarum DPPMA24W, DPPMASL33 *Lb. fermentum* DPPMA114 *Lactobacillus rhamnosus* DPPMAAZ1	30 °C for 96 h	Increase of isoflavone aglycones especially equol	[65]
Cowpeas	Spontaneously fermented *Lb. plantarum* ATCC 14917	37 °C for 48 h	Increase of phenolic acids derivatives and flavonoids. Improved antioxidant activity on DPPH	[66]
Lentils	*Lb. plantarum* CECT 748 and commercial protease	37 °C for 15 h	Reduction of ROS on RAW 264.7 cells	[67]
	Spontaneously fermented	35 °C for 4 days	Decrease of condensed tannin and increase of monomers.	[68]

4.1.1. Metabolic Activities Affecting Phenolic Acids

Phenolic acids are by far the most important food phenolics in terms of quantity, they represent one-third of dietary phenolics and can be present in soluble form within the cytoplasm or bounded to the cell wall [69]. Hydroxybenzoic and hydroxycinnamic acids may be decarboxylated by lactic acid bacteria (LAB) to the corresponding phenol or vinyl derivatives or hydrogenated by phenolic acid reductases [70]. Metabolites of phenolic acids conversion, compared to their precursors, have reduced antimicrobial activity [36] and it was also hypothesized that LAB use hydroxycinnamic acids as external acceptors of electrons, which allow them to gain one extra mole of ATP [71,72]. Strains of *Lactobacillus rossiae*, *Lactobacillus brevis*, and *Lactobacillus curvatus* have followed one of the two paths (decarboxylation or reduction), whereas strains of *Leuconostoc mesenteroides* and *Lactobacillus fermentum* were found not capable of metabolizing hydroxycinnamic acids [71]. Phenolic acids metabolism is influenced by the composition and intrinsic factors of the matrices, therefore, depending on the substrate, the metabolism can shift from decarboxylase to reductase [72]; nevertheless, the derivatives exert higher biological activities than their precursors [73].

Phenolic acid reductase and phenolic acid decarboxylase activities contributed to polyphenol metabolism in red sorghum fermented with *Lactobacillus plantarum* and *Lb. fermentum*. Ferulic acid was reduced to dihydroferulic acid, and caffeic acid was metabolized to vinylcatechol and ethylcatechol but also dihydrocaffeic acid [49]. Similar results were obtained during the fermentation of a malt-based beverage with a pool of LAB including *Lb. plantarum*, *Lb. brevis*, and *Lactobacillus amylolyticus* [51]. Savolainen et al. [40] studied the role of oxygen during the fermentation of a liquid wheat bran sourdough. It was observed that anaerobic conditions, in which lactic acid bacteria and endogenous heterotrophic bacteria grew better, induced the conversion of ferulic and caffeic acids into their corresponding derivatives, and increased the amount of sinapic acid. Aerobic conditions, which favored yeasts growth, was characterized by the presence of dihydroxyphenyl ethanol and hydroxyphenylacetaldehyde.

Phenolic acid metabolism was also reported in yeasts and fungi. Cinnamate caboxy-lyase activity, which transforms coumaric and ferulic acids into their vinyl derivatives, was reported in *S. cerevisiae* strains [74]. Species belonging to *Aspergillus*, *Fusarium*, and *Pycnoporus* genera as well as *Pseudomonas* were responsible for the decarboxylation of ferulic acid and eugenol and their further metabolization, through a lyase, to vanillin, vanillic, and protocatechuic acids [75,76].

As mentioned above, phenolic acids are often bound, as dimers, trimers and/or oligomers, to the plant cell wall polysaccharides such as xylan and pectin. Another class of enzymes involved in their metabolism is represented by feruloyl esterases which are capable of releasing ferulic acid and other cinnamic acid. Feruloyl esterases have been described in lactic acid bacteria, mostly *Lactobacillus plantarum* strains [69,77], in fungi of the genus *Aspergillus* and *Penicillium* [78,79], as well as in some *Bacillus*, *Pseudomonas*, and *Pseudoalteromonas* strains [77,79,80], and it was also hypothesized in *S. cerevisiae* [81]. Several authors studied the impact of bioprocessing with baker's yeast, LAB, and fungi, with or without the addition of commercial enzymes, on phenolic acids release in wheat bran [39–41,81], rye bran [46–48], rice bran [43,44], sorghum [49], and tef [51]. In all cases, substantial increases, especially of ferulic acid, were observed. Ferulic acid antioxidant properties are ascribed to its ability to inhibit lipid peroxidation and LDL oxidation greater than other hydroxycinnamic acids [82]. Anson et al. [38] fermented wheat bran with baker's yeast and used it to produce a fortified bread having ferulic, *p*-coumaric, and sinapic acids content up to three-fold higher than unprocessed bran. Breads were subjected to gastro-intestinal digestion in vitro, although phenolic compounds bioavailability substantial increased, most of them were recovered from the jejunal compartment; only a small part of them was further metabolized in the colon section. Slightly different results were obtained by Koistinen et al. [83], who fermented rye bran with baker's yeast. Despite the extensive phenolic acid release caused by the bioprocessing (up to 30-fold higher than unfermented sample), when subjected to in vitro colon model, no differences were observed among the thesis.

Among fermented legumes, phenolic acid decarboxylase and esterase activities were reported in fermented cowpeas [66], lentils [67,68], and chickpea [84]. Soy phenolic composition, on the other hand, has been extensively characterized [85], as well as its changes during fermentation with bacterial [55,59–61] and fungal strains [54,56,58,86]. Dueñas et al. [54] observed that despite hydroxycinnamic acids content was higher than hydroxybenzoic in unfermented soy flour, hydroxybenzoic acids significantly increased during fermentation, up to seven-fold when *Aspergillus oryzae* was used. On the other hand, *p*-hydroxyphenylacetic acid, which was not detected in raw flours, reached up to 30 μg/g after fermentation [54]. Riciputi et al. [62] also characterized the bound phenolic profile of soymilk fermented with *Lactobacillus casei* and *Lb. acidophilus* to prepare a fermented version of tofu. Bound phenolics were mostly phenolic acids, of which syringic represented more than 30%, followed by *p*-coumaric, ferulic, and *p*-cumaroyl-hexose derivatives, which all increased compared to soybean flour and traditional tofu [62].

4.1.2. Metabolic Activities Affecting Flavonoids

Flavonoids, the other big group of food phenolics, are often glycosylated and several enzymes, belonging to the class of hydrolases, are produced by a great number of microorganisms, both bacteria and fungi [57]. Glycosyl hydrolases convert flavonoid glycosides to the corresponding aglycones, which show higher bioactivity in humans than their precursor glycosides [60]. Glucosidase activity from *Lactobacillus* spp. was responsible for the reduction of flavonoids glycosides in red sorghum, which corresponded to the increase of the aglycones, taxifolin, eriodictyol, and naringenin [49]. Bhanja et al. [87] suggested that besides β-glucosidases, different carbohydrate cleaving enzymes among which amylase and xylanase are responsible for the release of phenolics during solid state fermentation of wheat by *A. oryzae* and *Aspergillus awamori*. Among flavonoids, soy isoflavones are the most extensively studied for their health benefits. As for most flavonoids, they normally occur as glucoside-bound moieties, yet it is the aglycone form that is metabolically active, showing higher antioxidant activity and being absorbed in the intestines faster than their glucoside bound forms [85]. Fermentation with *Lb. casei* and *Lb. acidophilus* enabled the increased of the aglycones/glycosylated ratio in fermented tofu more than 10- and five-fold higher compared to soybean flour and traditional tofu, respectively. Fermentation also increased the content of genistein and daidzein, the major isoflavones found in soybean [62]. Di Cagno et al. [65] fermented soymilk with a pool of LAB strains selected for the high β-glucosidase activity, registering the increase of daidzein, genistein, glycitein, and especially equol. β-glucosidase is the major enzyme involved in isoflavones release and its activity was responsible for the improvement of the scavenging activity on DPPH radical in many soy-products fermented by *Bacillus* spp. [55,85], *Aspergillus* spp. [54,56], *Rhizopus* spp. [54,86], *Lactobacillus* spp. [60,61,63], and *Lentinus edodes* [58]. Dueñas et al. [54] reported that fungal fermentation acted more intensively in releasing isoflavones aglycones compared to flavanones and flavonols. In addition to the β-glucosidase activity, McCue et al. [58] also ascribed the increase of phenolic compounds to a laccase, an enzyme involved in lignin biodegradation by white-rot fungi, suggesting the possibility of a direct activity on polymeric phenolic substrate. In a recent study, instead, the effect of fermentation on soy and chickpea isoflavones was evaluated after legumes seeds were germinated [63]. The authors evaluated the effect of different process parameters adopted during germination. An increase in isoflavones glycosides and aglycones was already observed on sprouts, however, after incubation with *Lb. casei*, especially when seeds were germinated with blue light, a substantial boost in their content was observed.

4.1.3. Metabolic Activities Affecting Tannins

Tannase, which specifically breaks the galloyl ester bonds of hydrolyzable tannins, was first reported for several fungal species of the genus *Aspergillus*. Over the last two decades, many bacterial species of the genera, including *Streptococcus*, *Lonepinella*, *Bacillus*, and *Lactobacillus*, have also been reported to possess tannase activity [88]. However, tannase acting on condensed tannins was also descripted [89]. An example of this enzymatic activity was reported during a spontaneous fermentation of lentils [68]. Eight catechin and epicatechin dimers, trimers and tetramers were identified in raw lentils. Their content was halved after a spontaneous fermentation with a consequent increase, of the monomeric forms. A similar outcome, was observed during the fermentation of a soy product with *Bacillus pumilus* [55], nevertheless, lactobacilli, and bifidobacteria from gut microbiota as well as a strain of *Lb. plantarum* isolated from cheese, were found to cleave the heterocyclic ring of monomeric flavan-3-ols, giving rise to 1-(3′,4′-dihydroxyphenyl)-3-(2″,4″,6″-trihydroxyphenyl)propan-2-ol (3,4-diHPP-2-ol) [90]. This compound, which can be further degraded by the gut microbiota in several other substances, was recently studied for its antioxidant properties [91]. The authors found that 3,4-diHPP-2-ol had scavenging activity on ABTS and DPPH radicals higher than its precursor, as well as than other phenolic compounds. In addition, 3,4-diHPP-2-ol also showed the ability to reduce Fe^{3+} to Fe^{2+}, higher than Trolox.

4.2. Release of Antioxidant Peptides

With the bioactive peptides crypted in the native protein, the active amino acid sequences need to be released through proteolytic microorganisms. Most of the studies involving the formation of bioactive peptides by fermentation are carried out by lactic acid bacteria which own a complex system of proteases and peptidases. Their proteolytic system consists of an extracellular proteinase, a transport system specific for small peptides, and a multitude of intracellular specific, generic, endo-, and eso-peptidases [92]. A pool of selected lactic acid bacteria (comprising *Lactobacillus alimentarius*, *Lb. brevis*, *Lb. sanfranciscensis*, and *Lactobacillus hilgardii*) was selected to ferment several cereal flours [32]. The highest antioxidant activity on DPPH and inhibition of linoleic acid autoxidation were found for fermented whole wheat, spelt, rye, and kamut. Twenty-five peptides (8-57 amino acid residues), identified by nano-liquid chromatography-electrospray ionization-tandem mass spectrometry (nano-LC-ESI-MS/MS), showed ex vivo antioxidant activity on mouse fibroblasts artificially subjected to oxidative stress [32]. Galli et al. [93] adopted the same experimental design selecting 23 out of 131 LAB strains to singly inoculate wheat flour doughs. The sourdough extracts reduced ROS formation in several cell lines (RAW 264.7 murine macrophage, murine H-end endothelium cells and Human intestinal Caco-2 cells). Another recent study exploited the effect of fermentation with *Lb. plantarum* and a commercial protease on lentils [67]. Due to the flavonoids liberated during the bioprocessing, a reduction of ROS on RAW 264.7 macrophage cell line was observed. Bioprocessed lentils were later subjected to simulated gastro-intestinal digestion and, since it was observed an increase in the inhibition of ROS formation, the authors suggested that the pH changes caused structure modifications of both phenolics and peptides, releasing even more bioactive sequences [67].

Twenty-seven selected LAB were also used to ferment quinoa flours and screened for the ability to improve antioxidant activity [33]. The highest scavenging activity on DPPH radical (more than 80%) was found when fermentation was carried out with *Lb. plantarum* T6B4, T6C16, and T0A10, whereas only *Lb. plantarum* T1B6, T6B4, and T0A10 showed antioxidant activity on ABTS radical. Since the last strain also enabled the highest inhibition of linoleic acid autoxidation, it was the only extract further subjected to the purification and characterization of antioxidant peptide fractions. Five peptides, with sizes from 5 to 9 amino acid residues, identified by nano-LC-ESI-MS/MS, showed antioxidant activity on human keratinocytes NCTC 2544 artificially subjected to oxidative stress and resulted resistant to further hydrolysis by digestive enzymes [33]. Ex vivo antioxidant properties of extracts of quinoa seed fermented with *Rhizopus oligosporus* were also evaluated by Matsuo [94]. The extracts increased both activities of superoxide dismutase and glutathione peroxidase ex vivo, therefore, they were used to feed rats for 13 days, confirming the results in vivo.

Several studies reported the ability of *Bacillus* sp., during soybean and wheat germ fermentation, to improve the in vitro antioxidant activity on DPPH and ABTS radicals. However, the activity was only ascribed to peptides; it was not demonstrated on purified fractions, nor were the sequences identified [95–98]. Conversely, rice proteins, residues from starch extraction, were hydrolyzed with a mixture of commercial proteolytic enzymes and *Bacillus pumilus* AG1, showing antioxidant activity towards ABTS radical. Almost all the peptides contained in the hydrolysate showed one or more features typical of well-known antioxidant peptides, most probably conferring a synergic antioxidant effect to the mixture with the potential to be used as functional ingredient [99].

Nevertheless, fungal eso- and endo-proteases play an important role in physiology and development of fungi, they are widely used in food industry [100], and examples of antioxidant peptides obtained by fungal fermentation were also reported (Table 2). Fermentation of soy flour with *Aspergillus oryzae* enabled the hydrolysis of native proteins into small molecular weight peptides, 90% of which were less than 3 kDa. The fermented soy was found to have antioxidant activity on DPPH radical higher than that of the positive controls (α-tocopherol and γ-oryzanol). It also inhibited 51.2% of linoleic acid oxidation, which was equivalent to 77% of the antioxidant activity of α-tocopherol [101].

As for phenolic compounds, several research papers focused on the antioxidant properties of soybean peptides as a result of fermentation with *Bacillus subtilis* [52,53], *Lb. plantarum* [102], *R. microspores* [103], *A. oryzae* [104], *Bifidobacterium* sp. [105]. Among the properties evaluated, the peptides possessed strong Cu^{2+} chelation ability, superoxide radical scavenging activity, reducing power potential and some of the peptides considered also resisted simulated gastro-intestinal digestion [53]. In one case, after assessing the radical scavenging activities on DPPH and ABTS, and the chelating ability of ferrous ions in vitro, the antioxidant properties of fermented soy were proven in vivo on rats, confirming the increase of superoxide dismutase activity in liver and kidney, and glutathione peroxidase activity in kidney [104]. A very recent study exploited the nutritional properties of a traditional soy-based Indonesian fermented food bought at local business with different level of sanitation [106]. The sanitation conditions clearly influence the microflora involved in the fermentation therefore affecting the formation of functional peptides. The identified peptides from the three different stores, differing in both number and molecular weight, showed similar and dissimilar features regarding amino acid sequences and functionalities. Yet, it was the cleanest production facility that had the highest number of peptides associated with functional properties including antioxidant activity [106]. Hence, this study enlightened the considerable impact of the starter in the success of the fermentation process.

Table 2. Main effect of bioactive peptides and amino acids derivatives on the antioxidant activity of fermented cereals, pseudocereals, and legumes.

Matrix	Microorganisms Employed	Process Parameters	Effect	Reference
Wheat flour	*Lactobacillus farciminis* A11, A19, H3, *Lactobacillus rossiae* A20, Gd40, *Lactobacillus sanfranciscensis* B3, I4, *Lactobacillus plantarum* O4, *Lactobacillus brevis* A7	30 °C for 24 h	Reduction of ROS on RAW 264.7, H-end and Caco-2 cells	[93]
Defatted wheat germ	*Bacillus subtilis* B1	37 °C for 24 h	Unidentified peptides	[96]
Rice protein	*Bacillus pumilus* AG1	37 °C for 72 h	Identified peptide sequences with high antioxidant activity	[99]
Wheat, spelt, rye, and kamut flours	Pool of selected LAB	37 °C for 24 h	Identified peptide sequences with high antioxidant activity	[32]
Quinoa	*Rhizopus oligosporus* NRRL2710	36 °C for 24 h	Increased ex vivo and in vivo activities of superoxide dismutase, CSHPx, and TBARS	[94]
	L. plantarum T0A10	37 °C for 24 h	Identified peptides with antioxidant activity on human keratinocytes NCTC 2544	[33]

Table 2. *Cont.*

Matrix	Microorganisms Employed	Process Parameters	Effect	Reference
Soy	*R. oligosporus* ATCC 64063 *Aspergillus oryzae* DSM 1861 *Neurospora intermedia* DSM 1965	31 °C for 6 days 25 °C for 6 days 30 °C for 5 days	Increase of OH and ABTS radical scavenging activity due to potentially bioactive peptides	[107]
	Spontaneously sourdough containing *Rhizopus* spp., LAB and yeasts	Room temperature for 2 days	Identified peptide sequences with potential antioxidant activity	[106]
	B. subtilis SHZ	30 °C for 36 h	Superoxide radical scavenging activity and reducing power potential	[52]
	B. subtilis MTCC5480, MTCC1747	42 °C for 24 h	Increase in the antioxidant activity due to both phenolic compounds and peptides	[53]
	Lb, plantarum Lp6	30 °C for 24 h	Identified peptide sequences with high antioxidant activity	[102]
	Aspergillus orizae	30 °C for 3 days 45 °C for 4 days	Unidentified low molecular weight peptides	[100]
	Rhizopus microsporus	36 °C for 25 h	Improvement in the antioxidant activity attributed to amino acids and peptides	[97]
	Bifidobacterium sp.	37 °C for 48 h	Inhibition of ascorbate autoxidation, superoxide radical scavenging activity and reducing power potential peroxide	[105]
	Aspergillus oryzae	30 °C for 60 h	Superoxide dismutase and glutathione peroxidase activities *in vivo*	[104]
Kidney beans	*B. subtilis* CECT 39T *Lb. plantarum* CECT 748T	30 °C for 96 h 37 °C for 96 h	Improved antioxidant activity on ORAC-FL	[108]
Lentils	*Lb. plantarum* CECT 748 and commercial protease	37 °C for 15 h	Reduction of ROS on RAW 264.7 cells	[67]

4.3. Secondary Effects of Fermentation

4.3.1. Vitamins

The ability of fermentation to modify phenolic and protein composition strictly relies on the metabolic activities of the specific starter used; however, the effect of acidification on endogenous proteinases or other enzymes involved in phenolics and protein metabolism cannot be excluded [69,92], as well as improving phenol solubility [49]. Few authors have also reported changes in vitamin content during fermentation [42,61,109] (Table 3). Vitamin E, also known as tocopherol, is an important antioxidant, which, due to a chroman group, halts lipid peroxidation by donating its phenolic hydrogen to peroxyl radicals forming tocopheroxyl radicals that are unreactive and unable to continue the oxidative chain reaction [1]. However, only photosynthetic microorganisms are known to accumulate detectable amounts of tocopherols; therefore, other factors participate to its increase during some food fermentations [110]. Fermented soy germ extracts exhibited a higher inhibition effect against the superoxide anion radical and lesser but significant ferric-reducing and DPPH radical scavenging effects compared to raw soy germ, which was ascribed to an increase in phenolic acids and isoflavones but also to tocopherols [61]. On the contrary, a decrease in tocopherol was observed in lupins fermentation, both spontaneous and inoculated with a strain of *Lb. plantarum* [111], Małgorzata et al. [112] instead, observed a decrease in the scavenging activity on ABTS of buckwheat fermented with *R. oligosporus*, despite the increase of tocopherol content. When *Cordyceps sinensis*, a fungus used in Chinese traditional medicine, was used to ferment stale rice, vitamin E concentration was doubled compared to unfermented rice and superoxide dismutase activity increased. Fermented rice was administrated to mice for 40 days inhibiting oxidative enzymes in brain and liver therefore delaying senescence [109]; and similar effects were also obtained by feeding rabbits with extracts of red bean fermented with *B. subtilis* [113]. To a combined action of polyphenols and vitamins was ascribed the protective effect against oxidative stress in mice after the consumption of bread made with Kamut sourdough [42,114]. However, Kamut breads, especially those with sourdough had a high content of selenium, which is known to participate in cells oxidative stress protection [114].

Table 3. Secondary effect of fermentation on the antioxidant activity of cereals and legumes.

Matrix	Microorganisms Employed	Process Parameters	Effect	Reference
Stale rice	*Cordyceps sinensis*	25 °C for 7 days	Increase of tocopherol and superoxide dismutase activity in mice	[109]
Rice medium	*Lactobacillus kefiranofaciens*		EPS responsible of atherosclerosis prevention	[115]
Wheat distillers' grains	*Preussia aemulans*	Room temperature for 7 days	EPS formation with high radical scavenging activity and metal ions chelating ability	[116]
Wheat germ and bran	Yeasts and lactobacilli		High vitamin content responsible for in vivo antioxidant activity in liver and kidney	[117]
Kidney beans	*Bacillus subtilis* IMR-NK1	30 °C for 48 h	Improved antioxidant activity on DPPH, reducing power potential and Fe^{2+} chelating ability	[118]
	Bacillus subtilis BCRC 14716	30 °C for 48 h	Increased vitamin E levels in liver and brain of rabbit, and superoxide dismutase activity in the brain	[113]

Great attention has gained a preparation known as Lisosan G®, a mineral- and vitamin-rich powder registered as nutritional supplement, which consists in wheat germ and bran fermented with a mix of lactobacilli and yeast strains [119]. Its antioxidant activity in vitro was confirmed after supplementation in the diet of 40 rabbits for 60 days. Lisosan G was found to reduce reactive oxygen metabolites and increase vitamin A and E concentrations in the blood. In addition, it caused the induction of antioxidant enzymes in the liver and kidney of the treated rabbits [117].

4.3.2. Production of Exopolysaccharides

Exopolysaccharides (EPS) are long-chain polysaccharides produced by microorganisms, as a response to environmental stress, using various sugars as substrates. EPS are either associated with cell surfaces or secreted into the environment and can be classified as homo- or hetero-exopolysaccharides, depending on the composing sugar units [120]. The EPS biosynthetic pathway is very complex and includes several enzymes and glucose-6-phosphate appears to be a key intermediate linking between the anabolic pathways of EPS production and the catabolic pathways of sugar degradation [121]. EPS can be produced by bacteria, yeasts, or filamentous fungi and their physiological role depends on the microorganism producing them [120]. EPS are mostly employed in food industry as texture modifiers; however, fungal and bacterial EPS have been proved to have anticancer, antimicrobial, hypocholesterolemic, hypoglycemic, and also antioxidant activity [122]. It was suggested that their radical scavenging ability is associated to the molecular weight and the number of hydroxyl and amino groups, yet the relationship between antioxidant activity and physico-chemical properties or structural features is still uncertain due to opposite results [122]. Many purified EPS produced during fermentation with species of the genera *Lactobacillus* [123–126], *Lactococcus* [127,128], *Cordyceps* [129,130], *Aspergillus* [131] were found to have antioxidant activity in vitro and in vivo. Scavenging activity towards DPPH, ABTS and OH radicals, metal ion chelating ability, and inhibition of linoleic acid peroxidation, as well as protective effect against Caco-2 cells oxidative stress and increased superoxide dismutase in mice are among the properties demonstrated; however, there are few studies that have confirmed it during food fermentation. A study conducted few years ago, explored the potential of wheat distillers' grains water extract fermentation to produce EPS by *Preussia aemulans* [116]. Compared to the unfermented extract, the fermented one had 36% of EPS higher and the scavenging activity against DPPH, ABTS, and OH radicals were assessed. One of the EPS fractions showed antioxidant activity comparable to that of ascorbic acid, as well as high ability to chelate metal ions and these properties were found to be dose-dependent [116]. Uchida et al. [115] studied the effect of rice kefiran, an EPS produced in a rice medium by *Lactobacillus kefiranofaciens*, on the diet of rabbits. The authors concluded that kefiran prevents the onset of atherosclerosis in hypercholesterolemic rabbits through antinflamatory and antioxidant actions [118].

5. Conclusions

The increasing interest in nutraceuticals reflects consumers' attention towards studies indicating that specific diets or components in the diet are associated with lower risk of certain disease. Consequentially, consumer trends have shifted towards super foods that not only fulfill basic nutritious requirement, but also exert any number of functional features while being natural without additives. The major role of antioxidant compounds in preserving food shelf life, as well as providing health promoting benefits, combined with the increasing concern towards synthetic antioxidants, has led many authors to look for natural ways of increasing their content through fermentation. Phenolic compounds are the most studied substances displaying such properties, nevertheless, also peptides and protein derivatives, vitamins, and EPS, released or produced by the complex microbial enzymatic system, have been proven to exert antioxidant activities in vitro, ex vivo and in vivo (Figure 1).

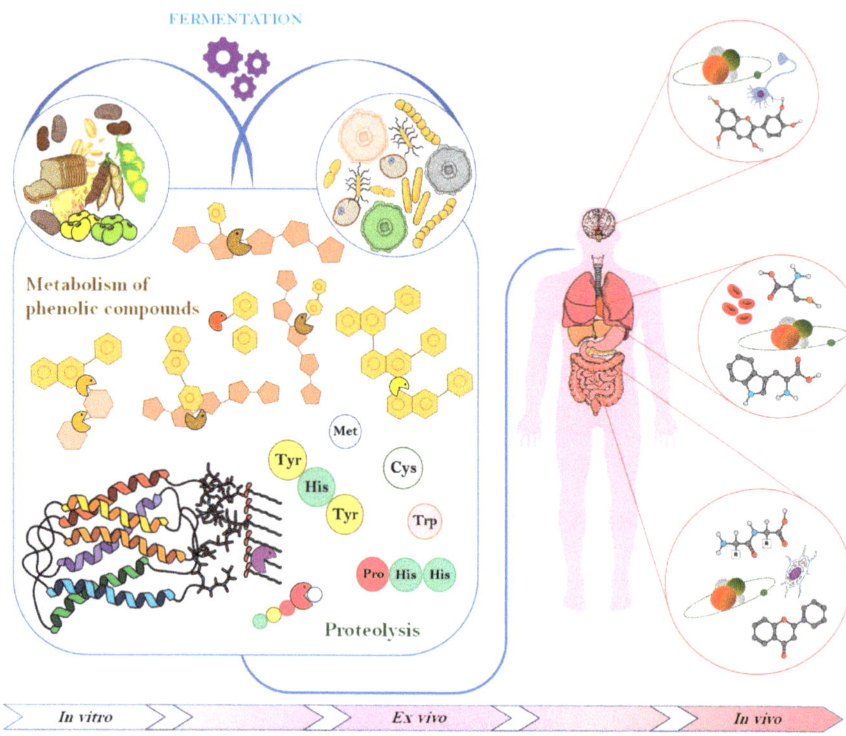

Figure 1. Schematic representation of the main effects of fermentation on food antioxidant compounds and consequent impact on human body.

Since most of the metabolic activities responsible for the antioxidant properties highlighted above are species- or strain- specific, starters selection is a pivotal step, and, even though the strains employed in these fermentations are recognized as safe, process parameters as well as many other factors, both affecting microbial and nutritional quality of fermented foods, need to be considered. Furthermore, as previously elucidated, an activity in vitro does not always correspond to an actual physiological function in vivo due to modifications occurring during the gastrointestinal digestion, therefore affecting the potential bioavailability of such compounds. Fermentation has the potential to meet consumers' requirements, yet more in vivo studies are needed.

Author Contributions: M.V. collected literature and wrote the manuscript, V.V., and C.G.R. reviewed and edited the manuscript. All the authors read and approved the manuscript.

Funding: The research was supported by the Spanish Ministry of Science, Innovation and Universities (project RTI2018-099835-A-I00).

Acknowledgments: V.V. thanks the Spanish Ministry of Economy and Competitiveness (MINECO) for "Ramon y Cajal" contract (RYC-2015-18795).

Conflicts of Interest: The authors declare no conflict of interest.

References

1. Carocho, M.; Ferreira, I.C. A review on antioxidants, prooxidants and related controversy: Natural and synthetic compounds, screening and analysis methodologies and future perspectives. *Food Chem. Toxicol.* **2013**, *51*, 15–25. [CrossRef] [PubMed]
2. Lobo, V.; Patil, A.; Phatak, A.; Chandra, N. Free radicals, antioxidants and functional foods: Impact on human health. *Pharmacogn. Rev.* **2010**, *4*, 118. [CrossRef] [PubMed]
3. Griffiths, K.; Aggarwal, B.; Singh, R.; Buttar, H.; Wilson, D.; De Meester, F. Food antioxidants and their anti-inflammatory properties: A potential role in cardiovascular diseases and cancer prevention. *Diseases* **2016**, *4*, 28. [CrossRef] [PubMed]
4. Alam, M.N.; Bristi, N.J.; Rafiquzzaman, M. Review on in vivo and in vitro methods evaluation of antioxidant activity. *Saudi Pharm. J.* **2013**, *21*, 143–152. [CrossRef] [PubMed]
5. Brewer, M.S. Natural antioxidants: Sources, compounds, mechanisms of action, and potential applications. *Compr. Rev. Food Sci. Food Saf.* **2011**, *10*, 221–247. [CrossRef]
6. Atmaca, G. Antioxidant effects of sulfur-containing amino acids. *Yonsei Med. J.* **2004**, *45*, 776–788. [CrossRef] [PubMed]
7. Rizzello, C.G.; Tagliazucchi, D.; Babini, E.; Rutella, G.S.; Saa, D.L.T.; Gianotti, A. Bioactive peptides from vegetable food matrices: Research trends and novel biotechnologies for synthesis and recovery. *J. Funct. Foods* **2016**, *27*, 549–569. [CrossRef]
8. Sarmadi, B.H.; Ismail, A. Antioxidative peptides from food proteins: A review. *Peptides* **2010**, *31*, 1949–1956. [CrossRef]
9. Pokorný, J. Are natural antioxidants better–and safer–than synthetic antioxidants? *Eur. J. Lipid Sci. Technol.* **2007**, *109*, 629–642. [CrossRef]
10. Box, J.D. Investigation of the Folin-Ciocalteau phenol reagent for the determination of polyphenolic substances in natural waters. *Water Res.* **1983**, *17*, 511–525. [CrossRef]
11. Shahidi, F.; Zhong, Y. Measurement of antioxidant activity. *J. Funct. Foods* **2015**, *18*, 757–781. [CrossRef]
12. Singh, B.; Singh, J.P.; Kaur, A.; Singh, N. Phenolic composition and antioxidant potential of grain legume seeds: A review. *Food Res. Int.* **2017**, *101*, 1–16. [CrossRef] [PubMed]
13. Gobbetti, M.; De Angelis, M.; Di Cagno, R.; Calasso, M.; Archetti, G.; Rizzello, C.G. Novel insights on the functional/nutritional features of the sourdough fermentation. *Int. J. Food Microbiol.* **2019**, *302*, 103–113. [CrossRef] [PubMed]
14. Verni, M.; Rizzello, C.G.; Coda, R. Fermentation biotechnology applied to cereal industry by-products: Nutritional and functional insights. *Front. Nutr.* **2019**, *6*, 42. [CrossRef] [PubMed]
15. Dey, T.B.; Chakraborty, S.; Jain, K.K.; Sharma, A.; Kuhad, R.C. Antioxidant phenolics and their microbial production by submerged and solid state fermentation process: A review. *Trends Food Sci. Technol.* **2016**, *53*, 60–74.

16. Manach, C.; Scalbert, A.; Morand, C.; Rémésy, C.; Jiménez, L. Polyphenols: Food sources and bioavailability. *Am. J. Clin. Nutr.* **2004**, *79*, 727–747. [CrossRef] [PubMed]
17. Lupea, A.X.; Pop, M.; Cacig, S. Structure-radical scavenging activity relationships of flavonoids from *Ziziphus* and *Hydrangea* extracts. *Rev. Chim.* **2008**, *59*, 309–313.
18. Shahidi, F.; Naczk, M. *Phenolics in Food and Nutraceuticals*; CRC press: Boca Raton, FL, USA, 2003; pp. 1–558.
19. Shahidi, F.; Ambigaipalan, P. Phenolics and polyphenolics in foods, beverages and spices: Antioxidant activity and health effects–A review. *J. Funct. Foods* **2015**, *18*, 820–897. [CrossRef]
20. Shahidi, F.; Zhong, Y. Bioactive peptides. *J. AOAC Int.* **2008**, *91*, 914–931.
21. Chen, H.M.; Muramoto, K.; Yamauchi, F.; Fujimoto, K.; Nokihara, K. Antioxidative properties of histidine-containing peptides designed from peptide fragments found in the digests of a soybean protein. *J. Agric. Food Chem.* **1998**, *46*, 49–53. [CrossRef]
22. Sánchez, A.; Vázquez, A. Bioactive peptides: A review. *Food Qual. Saf.* **2017**, *1*, 29–46. [CrossRef]
23. Nanditha, B.; Prabhasankar, P. Antioxidants in bakery products: A review. *Crit. Rev. Food Sci. Nutr.* **2008**, *49*, 1–27. [CrossRef]
24. EFSA. Panel on food additives and nutrient sources added to food (ANS); Scientific opinion on the reevaluation of butylated hydroxyanisole—BHA (E 320) as a food additive. *EFSA J.* **2011**, *9*, 2392. [CrossRef]
25. EFSA. Panel on food additives and nutrient sources added to food (ANS); Scientific opinion on the reevaluation of butylated hydroxytoluene BHT (E 321) as a food additive. *EFSA J.* **2012**, *10*, 2588. [CrossRef]
26. Carbonell-Capella, J.M.; Buniowska, M.; Barba, F.J.; Esteve, M.J.; Frígola, A. Analytical methods for determining bioavailability and bioaccessibility of bioactive compounds from fruits and vegetables: A review. *Compr. Rev. Food Sci. Food Saf.* **2014**, *13*, 155–171. [CrossRef]
27. Martin, K.R.; Appel, C.L. Polyphenols as dietary supplements: A double-edged sword. *Nutr. Diet. Suppl.* **2010**, *2*, 1–12. [CrossRef]
28. Hackman, R.M.; Polagruto, J.A.; Yan Zhu, Q.; Sun, B.; Fujii, H.; Keen, C.L. Flavanols: Digestion, absorption and bioactivity. *Phytochem. Rev.* **2008**, *7*, 195–208. [CrossRef]
29. D'Archivio, M.; Filesi, C.; Di Benedetto, R.; Gargiulo, R.; Giovannini, C.; Masella, R. Polyphenols, dietary sources and bioavailability. *Ann. Ist. Super. Sanita* **2007**, *43*, 348.
30. Cermak, R.; Durazzo, A.; Maiani, G.; Böhm, V.; Kammerer, D.R.; Carle, R.; Wiczkowski, W.; Piskula, M.K.; Galensa, R. The influence of postharvest processing and storage of foodstuffs on the bioavailability of flavonoids and phenolic acids. *Mol. Nutr. Food Res.* **2009**, *53*, S184–S193. [CrossRef]
31. García-Mora, P.; Martín-Martínez, M.; Bonache, M.A.; González-Múniz, R.; Peñas, E.; Frias, J.; Martinez-Villaluenga, C. Identification, functional gastrointestinal stability and molecular docking studies of lentil peptides with dual antioxidant and angiotensin I converting enzyme inhibitory activities. *Food Chem.* **2017**, *221*, 464–472. [CrossRef]
32. Coda, R.; Rizzello, C.G.; Pinto, D.; Gobbetti, M. Selected lactic acid bacteria synthesize antioxidant peptides during sourdough fermentation of cereal flours. *Appl. Environ. Microbiol.* **2012**, *78*, 1087–1096. [CrossRef]
33. Rizzello, C.G.; Lorusso, A.; Russo, V.; Pinto, D.; Marzani, B.; Gobbetti, M. Improving the antioxidant properties of quinoa flour through fermentation with selected autochthonous lactic acid bacteria. *Int. J. Food Microbiol.* **2017**, *241*, 252–261. [CrossRef]
34. Filannino, P.; Di Cagno, R.; Gobbetti, M. Metabolic and functional paths of lactic acid bacteria in plant foods: Get out of the labyrinth. *Curr. Opin. Biotechnol.* **2018**, *49*, 64–72. [CrossRef]
35. Jiménez, N.; Esteban-Torres, M.; Mancheño, J.M.; de las Rivas, B.; Muñoz, R. Tannin degradation by a novel tannase enzyme present in some *Lactobacillus plantarum* strains. *Appl. Environ. Microbiol.* **2014**, *80*, 2991–2997. [CrossRef]
36. Raccach, M. The antimicrobial activity of phenolic antioxidant in food: A review. *J. Food Saf.* **1984**, *6*, 141–170. [CrossRef]
37. Sánchez-Maldonado, A.F.; Schieber, A.; Gänzle, M.G. Structure—function relationships of the antibacterial activity of phenolic acids and their metabolism by lactic acid bacteria. *J. Appl. Microbiol.* **2011**, *111*, 1176–1184. [CrossRef]
38. Anson, N.M.; Selinheimo, E.; Havenaar, R.; Aura, A.M.; Mattila, I.; Lehtinen, P.; Bast, A.; Poutanen, K.; Haenen, G.R. Bioprocessing of wheat bran improves in vitro bioaccessibility and colonic metabolism of phenolic compounds. *J. Agric. Food Chem.* **2009**, *57*, 6148–6155. [CrossRef]

39. Manini, F.; Brasca, M.; Plumed-Ferrer, C.; Morandi, S.; Erba, D.; Casiraghi, M.C. Study of the chemical changes and evolution of microbiota during sourdoughlike fermentation of wheat bran. *Cereal Chem.* **2014**, *91*, 342–349. [CrossRef]
40. Savolainen, O.I.; Coda, R.; Suomi, K.; Katina, K.; Juvonen, R.; Hanhineva, K.; Poutanen, K. The role of oxygen in the liquid fermentation of wheat bran. *Food Chem.* **2014**, *153*, 424–431. [CrossRef]
41. Xie, C.Y.; Gu, Z.X.; You, X.; Liu, G.; Tan, Y.; Zhang, H. Screening of edible mushrooms for release of ferulic acid from wheat bran by fermentation. *Enz. Microb. Technol.* **2010**, *46*, 125–128. [CrossRef]
42. Benedetti, S.; Primiterra, M.; Tagliamonte, M.C.; Carnevali, A.; Gianotti, A.; Bordoni, A.; Canestrari, F. Counteraction of oxidative damage in the rat liver by an ancient grain (Kamut brand khorasan wheat). *Nutrition* **2012**, *28*, 436–441. [CrossRef]
43. Schmidt, C.G.; Gonçalves, L.M.; Prietto, L.; Hackbart, H.S.; Furlong, E.B. Antioxidant activity and enzyme inhibition of phenolic acids from fermented rice bran with fungus *Rizhopus oryzae*. *Food Chem.* **2014**, *146*, 371–377. [CrossRef]
44. Razak, D.L.A.; Rashid, N.Y.A.; Jamaluddin, A.; Sharifudin, S.A.; Long, K. Enhancement of phenolic acid content and antioxidant activity of rice bran fermented with *Rhizopus oligosporus* and *Monascus purpureus*. *Biocatal. Agric. Biotechnol.* **2015**, *4*, 33–38. [CrossRef]
45. Webber, D.M.; Hettiarachchy, N.S.; Li, R.; Horax, R.; Theivendran, S. Phenolic profile and antioxidant activity of extracts prepared from fermented heat-stabilized defatted rice bran. *J. Food Sci.* **2014**, *79*, H2383–H2391. [CrossRef]
46. Katina, K.; Laitila, A.; Juvonen, R.; Liukkonen, K.H.; Kariluoto, S.; Piironen, V.; Landberg, R.; Aman, P.; Poutanen, K. Bran fermentation as a means to enhance technological properties and bioactivity of rye. *Food Microbiol.* **2007**, *24*, 175–186. [CrossRef]
47. Koistinen, V.M.; Katina, K.; Nordlund, E.; Poutanen, K.; Hanhineva, K. Changes in the phytochemical profile of rye bran induced by enzymatic bioprocessing and sourdough fermentation. *Food Res. Intern.* **2016**, *89*, 1106–1115. [CrossRef]
48. Nordlund, E.; Katina, K.; Aura, A.M.; Poutanen, K. Changes in bran structure by bioprocessing with enzymes and yeast modifies the in vitro digestibility and fermentability of bran protein and dietary fibre complex. *J. Cereal Sci.* **2013**, *58*, 200–208. [CrossRef]
49. Svensson, L.; Sekwati-Monang, B.; Lutz, D.L.; Schieber, A.; Ganzle, M.G. Phenolic acids and flavonoids in nonfermented and fermented red sorghum (*Sorghum bicolor* (L.) *Moench*). *J. Agric. Food Chem.* **2010**, *58*, 9214–9220. [CrossRef]
50. Shumoy, H.; Gabaza, M.; Vandevelde, J.; Raes, K. Soluble and bound phenolic contents and antioxidant capacity of tef injera as affected by traditional fermentation. *J. Food Compos. Anal.* **2017**, *58*, 52–59. [CrossRef]
51. Dongmo, S.N.; Sacher, B.; Kollmannsberger, H.; Becker, T. Key volatile aroma compounds of lactic acid fermented malt based beverages–impact of lactic acid bacteria strains. *Food Chem.* **2017**, *229*, 565–573. [CrossRef]
52. Yu, B.; Lu, Z.X.; Bie, X.M.; Lu, F.X.; Huang, X.Q. Scavenging and anti-fatigue activity of fermented defatted soybean peptides. *Eur. Food Res. Technol.* **2008**, *226*, 415–421. [CrossRef]
53. Sanjukta, S.; Rai, A.K.; Muhammed, A.; Jeyaram, K.; Talukdar, N.C. Enhancement of antioxidant properties of two soybean varieties of Sikkim Himalayan region by proteolytic *Bacillus subtilis* fermentation. *J. Funct. Foods* **2015**, *14*, 650–658. [CrossRef]
54. Dueñas, M.; Hernández, T.; Lamparski, G.; Estrella, I.; Muñoz, R. Bioactive phenolic compounds of soybean (*Glycine max* cv. *Merit*): Modifications by different microbiological fermentations. *Pol. J. Food Nutr. Sci.* **2012**, *62*, 241–250.
55. Cho, K.M.; Hong, S.Y.; Math, R.K.; Lee, J.H.; Kambiranda, D.M.; Kim, J.M.; Islam, S.A.; Yun, M.G.; Cho, J.J.; Lim, W.J.; et al. Biotransformation of phenolics (isoflavones, flavanols and phenolic acids) during the fermentation of cheonggukjang by *Bacillus pumilus* HY1. *Food Chem.* **2009**, *114*, 413–419. [CrossRef]
56. Georgetti, S.R.; Vicentini, F.T.M.C.; Yokoyama, C.Y.; Borin, M.D.F.; Spadaro, A.C.C.; Fonseca, M.J.V. Enhanced in vitro and in vivo antioxidant activity and mobilization of free phenolic compounds of soybean flour fermented with different β-glucosidase-producing fungi. *J. Appl. Microbiol.* **2009**, *106*, 459–466. [CrossRef]
57. Avila, M.; Jaquet, M.; Moine, D.; Requena, T.; Pelaez, C.; Arigoni, F.; Jankovic, I. Physiological and biochemical characterization of the two α-L-rhamnosidases of *Lactobacillus plantarum* NCC245. *Microbiology* **2009**, *155*, 2739–2749. [CrossRef]

58. McCue, P.; Horii, A.; Shetty, K. Mobilization of phenolic antioxidants from defatted soybean powders by *Lentinus edodes* during solid-state bioprocessing is associated with enhanced production of laccase. *Innov. Food Sci. Emerg. Technol.* **2004**, *5*, 385–392. [CrossRef]
59. Hu, Y.; Ge, C.; Yuan, W.; Zhu, R.; Zhang, W.; Du, L.; Xue, J. Characterization of fermented black soybean natto inoculated with *Bacillus natto* during fermentation. *J. Sci. Food Agric.* **2010**, *90*, 1194–1202. [CrossRef]
60. Landete, J.M.; Hernández, T.; Robredo, S.; Duenas, M.; de las Rivas, B.; Estrella, I.; Munoz, R. Effect of soaking and fermentation on content of phenolic compounds of soybean (*Glycine max* cv. *Merit*) and mung beans (*Vigna radiata* L. *Wilczek*). *Int. J. Food Sci. Nutr.* **2015**, *66*, 203–209. [CrossRef]
61. Hubert, J.; Berger, M.; Nepveu, F.; Paul, F.; Daydé, J. Effects of fermentation on the phytochemical composition and antioxidant properties of soy germ. *Food Chem.* **2008**, *109*, 709–721. [CrossRef]
62. Riciputi, Y.; Serrazanetti, D.I.; Verardo, V.; Vannini, L.; Caboni, M.F.; Lanciotti, R. Effect of fermentation on the content of bioactive compounds in tofu-type products. *J. Funct. Foods* **2016**, *27*, 131–139. [CrossRef]
63. Budryn, G.; Klewicka, E.; Grzelczyk, J.; Gałązka-Czarnecka, I.; Mostowski, R. Lactic acid fermentation of legume seed sprouts as a method of increasing the content of isoflavones and reducing microbial contamination. *Food Chem.* **2019**, *285*, 478–484. [CrossRef]
64. Fiorda, F.A.; de Melo Pereira, G.V.; Thomaz-Soccol, V.; Medeiros, A.P.; Rakshit, S.K.; Soccol, C.R. Development of kefir-based probiotic beverages with DNA protection and antioxidant activities using soybean hydrolyzed extract, colostrum and honey. *LWT-Food Sci. Technol.* **2016**, *68*, 690–697. [CrossRef]
65. Di Cagno, R.; Mazzacane, F.; Rizzello, C.G.; Vincentini, O.; Silano, M.; Giuliani, G.; De Angelis, M.; Gobbetti, M. Synthesis of isoflavone aglycones and equol in soy milks fermented by food-related lactic acid bacteria and their effect on human intestinal Caco-2 cells. *J. Agric. Food Chem.* **2010**, *58*, 10338–10346. [CrossRef]
66. Dueñas, M.; Fernández, D.; Hernández, T.; Estrella, I.; Muñoz, R. Bioactive phenolic compounds of cowpeas (*Vigna sinensis* L). Modifications by fermentation with natural microflora and with *Lactobacillus plantarum* ATCC 14917. *J. Sci. Food Agric.* **2005**, *85*, 297–304. [CrossRef]
67. Bautista-Expósito, S.; Peñas, E.; Dueñas, M.; Silván, J.M.; Frias, J.; Martínez-Villaluenga, C. Individual contributions of Savinase and *Lactobacillus plantarum* to lentil functionalization during alkaline pH-controlled fermentation. *Food Chem.* **2018**, *257*, 341–349. [CrossRef]
68. Bartolomé, B.; Estrella, I.; Hernandez, T. Changes in phenolic compounds in lentils (*Lens culinaris*) during germination and fermentation. *Z. Lebensm Unters Forsch. A* **1997**, *205*, 290–294. [CrossRef]
69. Gänzle, M.G. Enzymatic and bacterial conversions during sourdough fermentation. *Food Microbiol.* **2014**, *37*, 2–10. [CrossRef]
70. Rodríguez, H.; Curiel, J.A.; Landete, J.M.; de las Rivas, B.; de Felipe, F.L.; Gómez-Cordovés, C.; Mancheño, J.M.; Muñoz, R. Food phenolics and lactic acid bacteria. *Int. J. Food Microbiol.* **2009**, *132*, 79–90. [CrossRef]
71. Filannino, P.; Gobbetti, M.; De Angelis, M.; Di Cagno, R. Hydroxycinnamic acids used as external acceptors of electrons: An energetic advantage for strictly heterofermentative lactic acid bacteria. *Appl. Environ. Microbiol.* **2014**, *80*, 7574–7582. [CrossRef]
72. Filannino, P.; Bai, Y.; Di Cagno, R.; Gobbetti, M.; Gänzle, M.G. Metabolism of phenolic compounds by *Lactobacillus* spp. during fermentation of cherry juice and broccoli puree. *Food Microbiol.* **2015**, *46*, 272–279. [CrossRef]
73. Senger, D.R.; Li, D.; Jaminet, S.C.; Cao, S. Activation of the Nrf2 cell defense pathway by ancient foods: Disease prevention by important molecules and microbes lost from the modern western diet. *PLoS ONE* **2016**, *11*, e0148042. [CrossRef]
74. Chatonnet, P.; Dubourdieu, D.; Boidron, J.N.; Lavigne, V. Synthesis of volatile phenols by *Saccharomyces cerevisiae* in wines. *J. Sci. Food Agric.* **1993**, *62*, 191–202. [CrossRef]
75. Gasson, M.J.; Kitamura, Y.; McLauchlan, W.R.; Narbad, A.; Parr, A.J.; Parsons, E.L.H.; Payne, J.; Rhodes, M.J.C.; Walton, N.J. Metabolism of ferulic acid to vanillin. A bacterial gene of the enoyl-SCoA hydratase/isomerase superfamily encodes an enzyme for the hydration and cleavage of a hydroxycinnamic acid SCoA thioester. *J. Biol. Chem.* **1998**, *273*, 4163–4170. [CrossRef]
76. Longo, M.A.; Sanromán, M.A. Production of food aroma compounds: Microbial and enzymatic methodologies. *Food Technol. Biotechnol.* **2006**, *44*, 335–353.
77. Donaghy, J.; Kelly, P.F.; McKay, A.M. Detection of ferulic acid esterase production by *Bacillus* spp. and lactobacilli. *Appl. Microbiol. Biotechnol.* **1998**, *50*, 257–260. [CrossRef]

78. Benoit, I.; Danchin, E.G.; Bleichrodt, R.J.; de Vries, R.P. Biotechnological applications and potential of fungal feruloyl esterases based on prevalence, classification and biochemical diversity. *Biotechnol. Lett.* **2008**, *30*, 387–396. [CrossRef]
79. Mathew, S.; Abraham, T.E. Ferulic acid: An antioxidant found naturally in plant cell walls and feruloyl esterases involved in its release and their applications. *Crit. Rev. Biotechnol.* **2004**, *24*, 59–83. [CrossRef]
80. Aurilia, V.; Parracino, A.; Saviano, M.; D'Auria, S. The psychrophilic bacterium *Pseudoalteromonas halosplanktis* TAC125 possesses a gene coding for a cold-adapted feruloyl esterase activity that shares homology with esterase enzymes from γ-proteobacteria and yeast. *Gene* **2007**, *397*, 51–57. [CrossRef]
81. Coghe, S.; Benoot, K.; Delvaux, F.; Vanderhaegen, B.; Delvaux, F.R. Ferulic acid release and 4-vinylguaiacol formation during brewing and fermentation: Indications for feruloyl esterase activity in *Saccharomyces cerevisiae*. *J. Agric. Food Chem.* **2004**, *52*, 602–608. [CrossRef]
82. Anson, N.M.; Hemery, Y.M.; Bast, A.; Haenen, G.R. Optimizing the bioactive potential of wheat bran by processing. *Food. Funct.* **2012**, *3*, 362–375. [CrossRef]
83. Koistinen, V.M.; Nordlund, E.; Katina, K.; Mattila, I.; Poutanen, K.; Hanhineva, K.; Aura, A.M. Effect of bioprocessing on the in vitro colonic microbial metabolism of phenolic acids from rye bran fortified breads. *J. Agric. Food Chem.* **2017**, *65*, 1854–1864. [CrossRef]
84. Xiao, Y.; Xing, G.; Rui, X.; Li, W.; Chen, X.; Jiang, M.; Dong, M. Enhancement of the antioxidant capacity of chickpeas by solid state fermentation with *Cordyceps militaris* SN-18. *J. Funct. Foods* **2014**, *10*, 210–222. [CrossRef]
85. McCue, P.; Shetty, K. Health benefits of soy isoflavonoids and strategies for enhancement: A review. *Crit. Rev. Food Sci. Nutr.* **2004**, *44*, 361–367. [CrossRef]
86. Randhir, R.; Vattem, D.; Shetty, K. Solid-state bioconversion of fava bean by *Rhizopus oligosporus* for enrichment of phenolic antioxidants and L-DOPA. *Inn. Food Sci. Emerg. Technol.* **2004**, *5*, 235–244. [CrossRef]
87. Bhanja, T.; Kumari, A.; Banerjee, R. Enrichment of phenolics and free radical scavenging property of wheat koji prepared with two filamentous fungi. *Bioresour. Technol.* **2009**, *100*, 2861–2866. [CrossRef]
88. Iwamoto, K.; Tsuruta, H.; Nishitaini, Y.; Osawa, R. Identification and cloning of a gene encoding tannase (tannin acylhydrolase) from *Lactobacillus plantarum* ATCC 14917T. *Syst. Appl. Microbiol.* **2008**, *31*, 269–277. [CrossRef]
89. Ramírez-Coronel, M.A.; Viniegra-Gonzalez, G.; Darvill, A.; Augur, C. A novel tannase from *Aspergillus niger* with β-glucosidase activity. *Microbiology* **2003**, *149*, 2941–2946. [CrossRef]
90. Sánchez-Patán, F.; Tabasco, R.; Monagas, M.; Requena, T.; Peláez, C.; Moreno-Arribas, M.V.; Bartolomé, B. Capability of *Lactobacillus plantarum* IFPL935 to catabolize flavan-3-ol compounds and complex phenolic extracts. *J. Agric. Food Chem.* **2012**, *60*, 7142–7151. [CrossRef]
91. Gleńsk, M.; Hurst, W.J.; Glinski, V.B.; Bednarski, M.; Gliński, G.A. Isolation of 1-(30,40-Dihydroxyphenyl)-3-(2″,4″,6″-trihydroxyphenyl)-propan-2-ol from grape seed extract and evaluation of its antioxidant and antispasmodic potential. *Molecules* **2019**, *24*, 2466. [CrossRef]
92. Gobbetti, M.; De Angelis, M.; Corsetti, A.; Di Cagno, R. Biochemistry and physiology of sourdough lactic acid bacteria. *Trends Food Sci. Technol.* **2005**, *16*, 57–69. [CrossRef]
93. Galli, V.; Mazzoli, L.; Luti, S.; Venturi, M.; Guerrini, S.; Paoli, P.; Vincenzini, M.; Granchi, L.; Pazzagli, L. Effect of selected strains of lactobacilli on the antioxidant and anti-inflammatory properties of sourdough. *Int. J. Food Microbiol.* **2018**, *286*, 55–65. [CrossRef]
94. Matsuo, M. In vivo antioxidant activity of methanol extract from quinoa fermented with *Rhizopus oligosporus*. *J. Nutr. Sci. Vitaminol.* **2005**, *51*, 449–452. [CrossRef]
95. Apostolidis, E.; Kwon, Y.I.; Ghaedian, R.; Shetty, K. Fermentation of milk and soymilk by *Lactobacillus bulgaricus* and *Lactobacillus acidophilus* enhances functionality for potential dietary management of hyperglycemia and hypertension. *Food Biotechnol.* **2007**, *21*, 217–236. [CrossRef]
96. Niu, L.Y.; Jiang, S.T.; Pan, L.J. Preparation and evaluation of antioxidant activities of peptides obtained from defatted wheat germ by fermentation. *J. Food Sci. Technol.* **2013**, *50*, 53–61. [CrossRef]
97. Watanabe, N.; Fujimoto, K.; Aoki, H. Antioxidant activities of the water-soluble fraction in tempeh-like fermented soybean (GABA-tempeh). *Int. J. Food Sci. Nutr.* **2007**, *58*, 577–587. [CrossRef]
98. Zhu, Y.P.; Fan, J.F.; Cheng, Y.Q.; Li, L.T. Improvement of the antioxidant activity of Chinese traditional fermented okara (Meitauza) using *Bacillus subtilis* B2. *Food Control* **2008**, *19*, 654–661. [CrossRef]

99. Dei Più, L.; Tassoni, A.; Serrazanetti, D.I.; Ferri, M.; Babini, E.; Tagliazucchi, D.; Gianotti, A. Exploitation of starch industry liquid by-product to produce bioactive peptides from rice hydrolyzed proteins. *Food Chem.* **2014**, *155*, 199–206. [CrossRef]
100. Yike, I. Fungal proteases and their pathophysiological effects. *Mycopathologia* **2011**, *171*, 299–323. [CrossRef]
101. Lee, J.S.; Rho, S.J.; Kim, Y.W.; Lee, K.W.; Lee, H.G. Evaluation of biological activities of the short-term fermented soybean extract. *Food Sci. Biotechnol.* **2013**, *22*, 973–978. [CrossRef]
102. Amadou, I.; Gbadamosi, O.S.; Shi, Y.; Kamara, M.T.; Jin, S. Identification of antioxidative peptides from *Lactobacillus plantarum* Lp6 fermented soybean protein meal. *Res. J. Microbiol.* **2010**, *5*, 372–380. [CrossRef]
103. Gibbs, B.F.; Zougman, A.; Masse, R.; Mulligan, C. Production and characterization of bioactive peptides from soy hydrolysate and soy-fermented food. *Food Res. Int.* **2004**, *37*, 123–131. [CrossRef]
104. Wang, D.; Wang, L.J.; Zhu, F.X.; Zhu, J.Y.; Chen, X.D.; Zou, L.; Saito, M. In vitro and in vivo studies on the antioxidant activities of the aqueous extracts of Douchi (a traditional Chinese salt-fermented soybean food). *Food Chem.* **2008**, *107*, 1421–1428. [CrossRef]
105. Tamam, B.; Syah, D.; Suhartono, M.T.; Kusuma, W.A.; Tachibana, S.; Lioe, H.N. Proteomic study of bioactive peptides from tempeh. *J. Biosci. Bioeng.* **2019**, in press. [CrossRef]
106. Starzyńska-Janiszewska, A.; Stodolak, B.; Gómez-Caravaca, A.M.; Mickowska, B.; Martin-Garcia, B.; Byczyński, Ł. Mould starter selection for extended solid-state fermentation of quinoa. *LWT-Food Sci. Technol.* **2019**, *99*, 231–237. [CrossRef]
107. Limón, R.I.; Peñas, E.; Torino, M.I.; Martínez-Villaluenga, C.; Dueñas, M.; Frias, J. Fermentation enhances the content of bioactive compounds in kidney bean extracts. *Food Chem.* **2015**, *172*, 343–352. [CrossRef]
108. Zhang, Z.; Lei, Z.; Yun, L.; Zhongzhi, L.; Chen, Y. Chemical composition and bioactivity changes in stale rice after fermentation with *Cordyceps sinensis*. *J. Biosci. Bioeng.* **2008**, *106*, 188–193. [CrossRef]
109. Ogbonna, J.C. Microbiological production of tocopherols: Current state and prospects. *Appl. Microbiol. Biotechnol.* **2009**, *84*, 217–225. [CrossRef]
110. Wang, Y.C.; Yu, R.C.; Chou, C.C. Antioxidative activities of soymilk fermented with lactic acid bacteria and bifidobacteria. *Food Microbiol.* **2006**, *23*, 128–135. [CrossRef]
111. Frias, J.; Miranda, M.L.; Doblado, R.; Vidal-Valverde, C. Effect of germination and fermentation on the antioxidant vitamin content and antioxidant capacity of *Lupinus albus* L. var. *Multolupa*. *Food Chem.* **2005**, *92*, 211–220. [CrossRef]
112. Małgorzata, W.; Joanna, H.; Konrad, P.M. Effect of solid-state fermentation with *Rhizopus oligosporus* on bioactive compounds and antioxidant capacity of raw and roasted buckwheat groats. *It. J. Food Sci.* **2015**, *27*, 424–431.
113. Chou, S.T.; Chung, Y.C.; Peng, H.Y.; Hsu, C.K. Improving antioxidant status in aged mice by 50% ethanol extract from red bean fermented by *Bacillus subtilis*. *J. Sci. Food Agric.* **2013**, *93*, 2562–2567. [CrossRef]
114. Gianotti, A.; Danesi, F.; Verardo, V.; Serrazanetti, D.I.; Valli, V.; Russo, A.; Riciputi, Y.; Tossani, N.; Caboni, M.F.; Guerzoni, M.E.; et al. Role of cereal type and processing in whole grain in vivo protection from oxidative stress. *Front. Biosci.* **2011**, *16*, 1609–1618. [CrossRef]
115. Uchida, M.; Ishii, I.; Inoue, C.; Akisato, Y.; Watanabe, K.; Hosoyama, S.; Toida, T.; Ariyoshi, N.; Kitada, M. Kefiran reduces atherosclerosis in rabbits fed a high cholesterol diet. *J. Atheroscler. Throm.* **2010**, 1006100247. [CrossRef]
116. Li, Y.T.; Meng, S.L.; Wang, L.B.; Wang, Y.P.; Zu, X.Y.; Yang, Y.N.; Zhang, Z.Y. Antioxidative activity of exopolysaccharide extract from fermented wheat distillers. *Adv. J. Food Sci. Technol.* **2014**, *6*, 1067–1075. [CrossRef]
117. Pozzo, L.; Vizzarri, F.; Ciardi, M.; Nardoia, M.; Palazzo, M.; Casamassima, D.; Longo, V. The effects of fermented wheat powder (Lisosan G) on the blood lipids and oxidative status of healthy rabbits. *Food Chem. Toxicol.* **2015**, *84*, 1–7. [CrossRef]
118. Chung, Y.C.; Chang, C.T.; Chao, W.W.; Lin, C.F.; Chou, S.T. Antioxidative activity and safety of the 50 ethanolic extract from red bean fermented by *Bacillus subtilis* IMR-NK1. *J. Agric. Food Chem.* **2002**, *50*, 2454–2458. [CrossRef]
119. La Marca, M.; Beffy, P.; Pugliese, A.; Longo, V. Fermented wheat powder induces the antioxidant and detoxifying system in primary rat hepatocytes. *PLoS ONE* **2013**, *8*, e83538. [CrossRef]
120. Donot, F.; Fontana, A.; Baccou, J.C.; Schorr-Galindo, S. Microbial exopolysaccharides: Main examples of synthesis, excretion, genetics and extraction. *Carbohydr. Polym.* **2012**, *87*, 951–962. [CrossRef]

121. Welman, A.D.; Maddox, I.S. Exopolysaccharides from lactic acid bacteria: Perspectives and challenges. *Trends Biotechnol.* **2003**, *21*, 269–274. [CrossRef]
122. Wang, J.; Hu, S.; Nie, S.; Yu, Q.; Xie, M. Reviews on mechanisms of in vitro antioxidant activity of polysaccharides. *Oxid. Med. Cell. Longev.* **2016**. [CrossRef]
123. Adesulu-Dahunsi, A.T.; Jeyaram, K.; Sanni, A.I.; Banwo, K. Production of exopolysaccharide by strains of *Lactobacillus plantarum* YO175 and OF101 isolated from traditional fermented cereal beverage. *PeerJ* **2018**, *6*, e5326. [CrossRef]
124. Liu, C.F.; Tseng, K.C.; Chiang, S.S.; Lee, B.H.; Hsu, W.H.; Pan, T.M. Immunomodulatory and antioxidant potential of *Lactobacillus* exopolysaccharides. *J. Sci. Food Agric.* **2011**, *91*, 2284–2291. [CrossRef]
125. Patten, D.A.; Laws, A.P. *Lactobacillus*-produced exopolysaccharides and their potential health benefits: A review. *Benef. Microbes* **2015**, *6*, 457–471. [CrossRef]
126. Zhang, L.; Liu, C.; Li, D.; Zhao, Y.; Zhang, X.; Zeng, X.; Yang, Z.; Li, S. Antioxidant activity of an exopolysaccharide isolated from *Lactobacillus plantarum* C88. *Int. J. Biol. Macromol.* **2013**, *54*, 270–275. [CrossRef]
127. Guo, Y.; Pan, D.; Li, H.; Sun, Y.; Zeng, X.; Yan, B. Antioxidant and immunomodulatory activity of selenium exopolysaccharide produced by *Lactococcus lactis* subsp. *lactis*. *Food Chem.* **2013**, *138*, 84–89. [CrossRef]
128. Pan, D.; Mei, X. Antioxidant activity of an exopolysaccharide purified from *Lactococcus lactis* subsp. *lactis* 12. *Carbohydr. Polym.* **2010**, *80*, 908–914. [CrossRef]
129. Leung, P.H.; Zhao, S.; Ho, K.P.; Wu, J.Y. Chemical properties and antioxidant activity of exopolysaccharides from mycelial culture of *Cordyceps sinensis* fungus Cs-HK1. *Food Chem.* **2009**, *114*, 1251–1256. [CrossRef]
130. Yan, J.K.; Li, L.; Wang, Z.M.; Leung, P.H.; Wang, W.Q.; Wu, J.Y. Acidic degradation and enhanced antioxidant activities of exopolysaccharides from *Cordyceps sinensis* mycelial culture. *Food Chem.* **2009**, *117*, 641–646. [CrossRef]
131. Chen, Y.; Mao, W.; Tao, H.; Zhu, W.; Qi, X.; Chen, Y.; Li, H.; Zhao, C.; Yang, Y.; Hou, Y.; et al. Structural characterization and antioxidant properties of an exopolysaccharide produced by the mangrove endophytic fungus *Aspergillus* sp. Y16. *Biores. Technol.* **2011**, *102*, 8179–8184. [CrossRef]

foods

MDPI

Article

Isolation of *Penicillium citrinum* from Roots of *Clerodendron cyrtophyllum* and Application in Biosynthesis of Aglycone Isoflavones from Soybean Waste Fermentation

Duy Tien Doan [1], Duc Phuong Luu [1], Thanh Duong Nguyen [1], Bich Hoang Thi [2], Hong Minh Pham Thi [2,3], Huu Nghi Do [2,3], Van Huyen Luu [4], The Dan Pham [2,5], Van Thai Than [6], Hai Ha Pham Thi [7,*], Minh Quan Pham [2,3] and Quoc Toan Tran [2,3,*]

1 Institute of Chemistry, Vietnam Academy of Science and Technology, 18 Hoang Quoc Viet St., Cau Giay Dist., Hanoi 10072, Vietnam; doanduytien@yahoo.com (D.T.D.); luuphuongtheday@gmail.com (D.P.L.); ntduong182@gmail.com (T.D.N.)
2 Institute of Natural Products Chemistry, Vietnam Academy of Science and Technology, 18 Hoang Quoc Viet St., Cau Giay Dist., Hanoi 10072, Vietnam; bichhoang.inpc@gmail.com (B.H.T.); minhhcsh@gmail.com (H.M.P.T.); dohnghi@gmail.com (H.N.D.); danpt.bi9064@st.usth.edu.vn (T.D.P.); minhquanaries@gmail.com (M.Q.P.)
3 Graduate University of Science and Technology, Vietnam Academy of Science and Technology, Hanoi 10072, Vietnam
4 Hanoi University of Natural Resources and Environment, Hanoi 10072, Vietnam; lvhuyen@hunre.edu.vn
5 Vietnam Academy of Science and Technology, University of Science and Technology of Hanoi, Hanoi 10072, Vietnam
6 NTT Institute of High Technology, Nguyen Tat Thanh University, Ho Chi Minh City 700000, Vietnam; tvthai@ntt.edu.vn
7 Faculty of Biotechnology, Nguyen Tat Thanh University, Ho Chi Minh City 700000, Vietnam
* Correspondence: pthha@ntt.edu.vn (H.H.P.T.); tranquoctoan2010@gmail.com (Q.T.T.)

Received: 16 October 2019; Accepted: 28 October 2019; Published: 6 November 2019

Abstract: Soybeans offer an abundant source of isoflavones, which confer useful bioactivities when existing in aglycone forms. The conversion of isoflavones into aglycones via fermentation of soybean products is often realized by β-glucosidase, an enzyme produced by fungi. In this study, a filamentous fungus, *Clerodendron cyrtophyllum*, was isolated from root of *Clerodendron cyrtophyllum* Turcz, which was able to produce the highest activity of β-glucosidase up to 33.72 U/mL at 144 h during fermentation on Potato Dextrose Broth (PDB). The obtained fungus was grown on isoflavones-rich soybean extract to produce genistein and daidzein, achieving the conversion rate of 98.7%. Genistein and daidzein were isolated and purified by column chromatography using hexane/acetone (29:1/1:1), reaching purities of over 90% of total isoflavones, as identified and determined by TLC, LC-MS/MS, and ^{1}H and ^{13}C NMR spectroscopy. These results imply that the isolated *P. citrinum* is a potential fungal strain for industrial-scale production of genistein and daidzein from isoflavones-containing soybean extracts. These products may serve as potential raw materials for manufacture of functional foods that are based on aglycones.

Keywords: *Penicillium citrinum*; β-glucosidase; soybean extract; fermentation; isoflavones; aglycones

1. Introduction

Isoflavones are polyphenolics that exert estrogen-like effects and have been widely utilized in manufacture of foods and cosmetics [1]. Isoflavones exhibit a myriad of bioactivities including protecting against colon cancer and aging skin, easing postmenopausal symtoms in women, reducing the risk of osteoporosis, preventing cardiovascular disease, having antimutagenic, antioxidant, and anti-inflammatory effects, and acting as tyrosine kinase enzyme inhibitors [1,2]. Since the intestinal absorption of isoflavone in aglycone form, particularly geistein, is more expedited than in the glycosylated form, most beneficial biological properties have been suggested to originate from aglycone isoflavone [2]. However, the content of aglycones is significantly lower than that of the glycoside counterpart, accounting for only 5% of total isoflavones. Structurally, the phenolic group of isoflavones is bonded to glycosides and primary hydroxyl group of glucose moiety is bonded to 6-O-acetyl or 6-O-malonyl derivatives. It has been shown that isoflavones existed in 12 isomers: three free isoflavones (aglycones) and nine linked isoflavones (glucosides) [3–6].

In traditional Asian diets, the main supply of isoflavones is from soybean (*Glycine max*) [7]. Soybean contains a large amount of bioactive subtances such as isoflavones, saponin, phytosterol, protease inhibitor, inositol hexaphosphate, and trypsin inhibitors [8]. As the bioavailability of soy isoflavone aglycones is superior to that of the glycoside forms, the isolation of isoflavone aglycones from soy products as well as production of aglycones for pharmaceutical and functional food applications have attracted research interest recently [9,10].

Industrial production of soybean oil generates byproducts such as soapstocks and soy cakes. While the former, which is rich in fatty acids [11], could be utilized in production of animal feeds, soap, nonpetroleum waxes, and biofuels, the latter was recently suggested to play an important part in the biotechnological production of several high-added value chemicals, such as fumaric acid by fungi, carotenoids by yeasts, and polyol esters and wax esters by the enzymatic synthesis [12–14]. The production of such soybean-derived products are realized via various techniques including saponification [15], esterification, and enzymatic hydrolysis [16].

The common production technique of aglycones involves hydrolysis of isoflavone glucosides using chemical catalysts (e.g., bases or acids) [8]. Alternatively, the transformation of isoflavones glycosides to aglycones can be also achieved during soybean germination as a natural occurence or by selective enzymatic hydrolysis and fermentation with β-glucosidase-producing microorganisms [17–21]. The latter has attracted considerable attention by researchers as a green approach.

β-Glucosidase (β β -D-glucoside glucohydrolase; EC 3.2.1.21), an exocellulase responsible for the hydrolysis of the O-glycosyl linkage of terminal nonreducing β-D-glucosyl residues releasing β-D-glucose, e.g., the bond in cellobiose, has been the enzyme of interest in large scale production via fermentation with microbes [22,23]. To be specific, feasible microbes that have been utilized in β-glucosidase production include *Aspergillus* niger [24,25], *A. oryzae* [26], *Penicillium* brasilianum [27], *Phanerochaete chrysosporium* [28], *Aspergillus* oryzae [29], *Thermofilum pendens* [30], yeasts (mostly *Candida* sp.) and several bacterial species. Even though fermentation to produce β-glucosidase for hydrolysis using *A. niger*, a filamentous fungus, is popular and has been the standard in commercial β-glucosidase production [31], emerging approaches utilize various strains from the *Penicillium* genus [27]. These new techniques also produce lipase and esterase, which act as catalysts for hydrolysis of ester groups of 6-O-acetyl and 6-O-malonyl isoflavones derivatives for activation of hydrolysis of glycosides by β-glucosidase.

In this work, multiple β-glucosidases-producing fungi were isolated and screened from roots of the *Clerodendron cyrtophyllum* Turcz plant. The fungal strain that exhibited the highest β-glucosidase activity was selected and then used for fermentation of isoflavones-rich soybean extract to produce aglycones, genistein, and daidzein. The products were identified and determined by thin layer chromatography (TLC), liquid chromatography-tandem mass spectrometry (LC-MS/MS), and nuclear magnetic resonance (NMR).

2. Materials and Methods

2.1. Materials

Sabouraud-2% dextrose Broth (SDB), Sabouraud-4% glucose-Agar (SA), potato dextrose broth (PDB), potato dextrose agar (PDA), p-nitrophenyl-β-D-glucopyranoside (pNPG), p-nitrophenol (pNP), and methylene blue were purchased from Merck KGaA (Darmstadt, Germany). Roots of *Clerodendron cyrtophyllum* Turcz, locally collected at Hoai Duc ward, Hanoi, Vietnam, were coarsely ground to the size of 5 mm and then were divided equally to two parts. One part was subjected to sun drying and sterilization using UV light for 10 min while the other part was stored in gunny bags. The sized *C. cyrtophyllum* Turcz's roots were used for isolation of the β-glucosidase-producing fungi. Oil-extracted soybean residue (dry form) (*Glycine max* L.) was purchased from T&H Agriculture and Technology, JSC (Hanoi, Vietnam).

2.2. Isolation of Microorganisms

Fungal strains were isolated by serial dilution [32]. Briefly, 1 g of the sized *C. cyrtophyllum* Turcz's roots was mixed with 10 mL of sterile deionized water. The following serial dilution was performed to 10^{-5} and 0.1 mL of the suspension was dispersed onto the sterilized PDA agar plates, which were allowed to stand for 72 h at 30 °C to promote fungal growth. Different types of fungi were isolated. They were then subcultured on sterilized PDA plates several times to obtain pure cultures, which were maintained at 4 °C for further study.

2.3. Fungal Identification

The fungal cultures were obtained by repeatedly transferring onto sterilized PDA plates and inoculated at 30 °C for 72 h. After inoculation, characteristics such as color and size of colonies during the growth was monitored and recorded. A little amount of mycelial mat was placed on clean glass slide and then stained with methylene blue. These slides were microscopically analyzed for morphological characteristics. Fungal isolates were identified by comparing these characteristics with those listed in standard reference book [33].

To prepare DNA, each fungal strain was cultured and incubated on PDA plates for three days at 30 °C. Purelink Genomic DNA Kits (Life Technology, ThermoFisher Scientific, Massachusetts, USA) was used to obtain DNA from the hyphae of the isolate. Two primer pairs, ITS1 (5′-TCCGTAGGTGAACCTGCGG-3′) and ITS4 (5′-TCCTCCGCTTATTGATATGC-3′) were used to perform the amplification of the the internal transcribed spacer region (ITS) of the nuclear ribosomal DNA operon [34]. PCR experiments were performed as follows. First, a 20 μL mixture containing 5–50 ng of DNA, AccuPrep PCR premix (Bioneer, Daejeon, South Korea), and 5 pmol of each primer was created. The mixture was first denatured at 95 °C for 5 min. Afterwards, a total of 30 cycles of denaturation, annealing, and extension took place with the elapsed time of 30 s in each cycle. Temperature for the three processes was 95, 48, and 72 °C, respectively. The protocol concluded with extension at 72 °C for 7 min. Electrophoretic characterization of the obtained PCR products was performed in 1% agarose gel. Obtained genomic sequences were referenced to GenBank via BLAST search tool (http://blast.ncbi.nlm.nih.gov/Blast.cgi) and were found to be in line with reported sequences.

2.4. Enzymatic Activity

Fungal strains were maintained on PDA at 30 °C with periodic transfer of five days. Inoculum was prepared by growing the fungal mycelium in 100 mL sterilized basal medium, which consisted of 2 g/L $(NH_4)_2SO_4$, 1 g/L KH_2PO_4, 0.4 g/L $MgSO_4{\cdot}7H_2O$, and 0.5 g/L $CaCl_2$ (pH = 5.6) and 10% (*w*/*v*) PDB as carbon source in 250 mL Erlenmeyer conical flasks. Incubation was carried out at 120 rpm using a rotary shaker, at 30 °C for 72 h. After incubation, β-glucosidases produced by different fungi strains in flasks were extracted by supplementing with potassium phosphate buffer 20 mM (pH = 6), followed by shaking at 200 rpm for 1 h and then filted by using membrane Supro 450 size 0.45 um (Pall, Ann Arbor, MI, USA). The obtained filtrates were used as crude enzymes. The hydrolysis of p-nitrophenyl-β-D-glucopyranoside (pNPβG) substrate was taken as the measure for β-glucosidase activity of fungi [35]. The reaction mixture contained 30 μL of appropriately aliquot (the filtrate) from incubated flask, 60 μL of pNPβG and 210 μL of acetate buffer. The reaction commenced at 50 °C for 60 min, followed by addition of 300 μL of 1 M Na_2CO_3. The resultant release of p-nitrophenol (pNP) caused color change, which was measured spectrophotometrically at 410 nm. The enzyme quantity required to maintain the pNP release rate of 1 μmol per minute is equivalent to one unit of β-glucosidase activity.

2.5. Optimization of the β-Glucosidase Production

The incubation of *Penicillium citrinum* was carried out in a 250 mL Erlenmeyer flask containing 100 mL of described basal medium under rotary shaking at 120 rpm at 30 °C for seven days. β-glucosidase activity was determined at different intervals starting from the beginning of incubation, including 24, 48, 72, 96, 120, 144, and 168 h, using the aforementioned procedure.

Effect of carbon sources on growth and enzyme production were evaluated by the following procedure. First, 1 mL aliquot of SDP-grown *Penicillium* sp. was placed into a 250 mL Erlenmeyer flasks containing 100 mL the basal medium with adding 0.75% $(NH_4)_2SO_4$, 50 mM KH_2PO_4. Afterwards, different carbon sources including either PDB, corn, potato, barley, and Czapek's Dox (Cz) were supplemented at the concentration of 3% (*w*/*v*). Incubation of flasks was carried out using a rotary shaker (120 rpm) at 30 °C for the optimum time obtained in the previous section.

2.6. Fermentation of Soybean Residual Using P. citrinium

First, 50 g of soybean residue collected after the extraction process of soybean oil, accruing from the extraction process of soybean oil, was used as a feedstock for isoflavones extraction by using 100 mL of ethanol/water (70/30, *v*/*v*) and 70 °C for 2 h. The extracted mixture was filtered to obtain filtrate. The extraction experiment was repeated three times and the filtrates were pooled together before ethanol removal by vacuum distillation to obtain the final extract (designated as CDN) containing concentrated isoflavones for fermentation experiments. The fermentation mixture was made up by mixing 20% (*w*/*v*) glucose and 5–20% (*v*/*v*) of CDN extract, 0.75% (*w*/*v*) $(NH_4)_2SO_4$, 50 mM KH_2PO_4, pH 4.5 followed by sterilization at 121 °C for 20 min. A fungal amount of 2% (*v*/*v*) of *P. citrinium* was used to inoculate the medium prepared in 250 mL Erlenmeyer flasks and incubation at 30 °C for 120 h with shaking rate of 180 rpm. Samples were regularly taken at 24, 48, 72, 96, and 120 h of fermentation, followed by addition of ethanol, vortexing, and centrifugation to obtain supernatants for the determination of the conversion of glucosides, genistin, and daidzin into aglycones and corresponding glycosides.

Hydrolysis rate was calculated by the following equation:

$$H = \frac{\text{isoflavone glycosides in control samples} - \text{isoflavone glycosides in hydrolyzed samples}}{\text{isoflavone glycosides in control samples}} \times 100$$

Isoflavone aglycones in the hydrolyzed mixture were isolated and purified by means of column chromatography (Mini-C, diameter of 70 mm) on silica gel (Merck 60, 15–40 μm, height loaded of 10 cm) using hexane/acetone (29:1/1:1).

2.7. Determination of Glucosides and Aglycones Using TLC, LC-MS/MS, NMR

To perform TLC (thin layer chromatography) analysis, silica gel 60 F_{254} plates were used. The employed developing solvent was a mixture of chloroform/methanol/water (80:20:2, *v*/*v*/*v*, lower phase). The revelation of spots were performed by spraying with $FeCl_3$ 5% or 1% $Ce(SO_4)_2/H_2SO_4$, followed by exposure to UV lamp with wavelength of 254 nm till spots were clearly observed. Identification of glucosides and aglycones was made by comparison with reference standards. An LC instrument (Agilent 1100 system, Santa Clara, CA, USA) was employed to perform LC analysis at a detection wavelength of 260 nm, the injection volume of 20 μL with the reverse phase C25 ODS-2 (250 mm × 4.6 mm, 5 μm) column. Two solvents, A (acetonitrile 100%) and B (formic acid 0.15%), were used as the mobile phase. The gradient conditions were as follows. [100–50% B, 0–50 min]; [50–20% B, 50–52 min]; [20–0% B, 52–60 min] with a flow rate of 1.0 mL/min. Mass spectra of Electron Spray Ionisation (ESI) were recorded on the Agilent 1100 Series mass spectrometer connected with Varian 320-MS (LabWrench, Midland,

Canada). ^{1}H-NMR spectra were recorded by 500 MHz (Bruker XL-500, Billerica, MA, USA) with DMSO-d_6 or acetone-d_6 as solvents and tetramethylsilane as an internal standard. ^{13}C-NMR spectra were recorded at 125 MHz (Bruker XL-500) with DMSO-d_6 or acetone-d_6 as solvents and as internal standard.

2.8. Statistical Anaylsis

Experiments were carried out in triplicate for the accuracy of data. Statistical significant differences were realized at $p < 0.05$ via Student's *t*-test. Statistical analysis was performed in the JMP Pro 13.2 software.

3. Results

3.1. Isolation and Screening of β-Glucosidase-Producing Microorganisms

Roots from *Clerodendron cyrtophyllum* Turcz sp. were used as microbial sources to isolate fungi with β-glucosidase activity on PDA medium. At 10^{-5} dilution, colonies were isolated separately. They were then subcultured on sterilized PDA plates several times to obtain pure culture for identification and β-glucosidase enzyme assay. After several subcultures at 30 °C for 72 h, 10 fungi were isolated with different morphological characteristics of colony. Five fungi were isolated from fresh samples (designated as BC) of *C. cyrtophyllum* Turcz's while five other fungi were isolated from dried plant samples (designated as BSD).

For enzymatic screening and fungal selection, the ten fungal isolates were tested for their capability of enzyme production. The β-glucosidase activity of these fungi is shown in the Figure 1. The strains BC2 and BSD5 were determined as the most potential fungi for enzyme production (5.41 ± 0.27 and 8.53 ± 0.20 U/mL, respectively) in comparison with other strains (<0.63 U/mL) and were therefore selected for further studies.

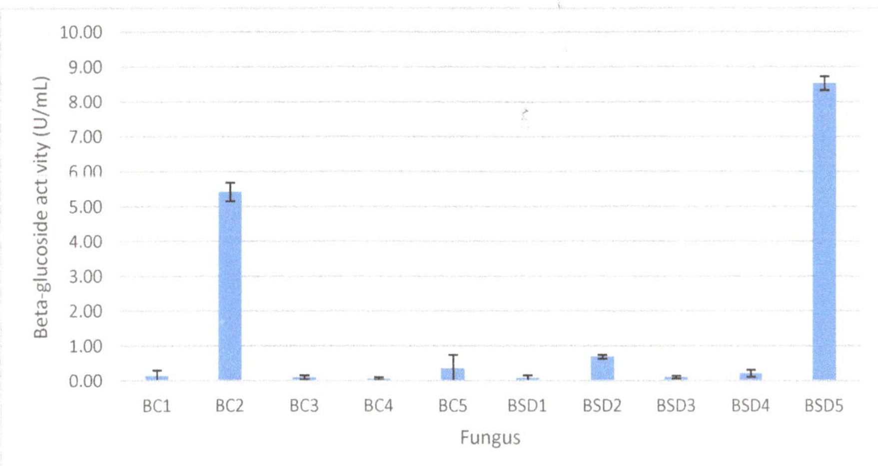

Figure 1. β-glucosidase activity of the isolated fungi.

3.2. Isolation and Screening of β-Glucosidase-Producing Microorganisms

On SA medium, colonies of BC2 grew fast and displayed a compact green or yellow basal felt enclosed by a layer of white of erect conidiophores. The diameter of colonies was approximately measured as 2.4 cm. Microscopically, conidiophore stipes are long and smooth-walled and the color of hyaline turned dark towards the vesicle. Conidial heads were biseriate, large, globose, and showed tendency to radiate, splitting into several loose columns with age. Additionally, phialides, usually in the form of septate metulae, were also observed. These morphological characteristics affirm that this fungus belongs to the *Aspergillus* species.

The shade of colonies of BSD5 growing on SA medium was green in color and a dense felt of conidiophores was observed. The average diameter of colonies after three days of incubation was 1.2 cm. Observing under a microscope, conidiophores were hyaline and smooth-walled. In addition, terminal verticils, carrying 3–5 metulae each, were observed on conidiophores. In each metula, around 3–7 phialides were recognized. Regarding conidia, they are smooth-walled and their shape appeared to be globose or subglobose. In addition, the production of conidia was basipetal from the phialides. Therefore, the described fungus belonged to the *Penicillium* species.

After identification by morphological characteristics, genomic DNA of the *Penicillium* sp. was identified to ensure that this fungus belonged to *Penicillium* sp. Figure 2 shows the PCR products on agarose gel 1% in which lane 1 is marker proteins and lane 2 is DNA of the fungus. It can be demonstrated that the molecular mass of this fungus is 540 bp. By using ITS method, the genome DNA of the *Penicillium* sp. was sequenced as:

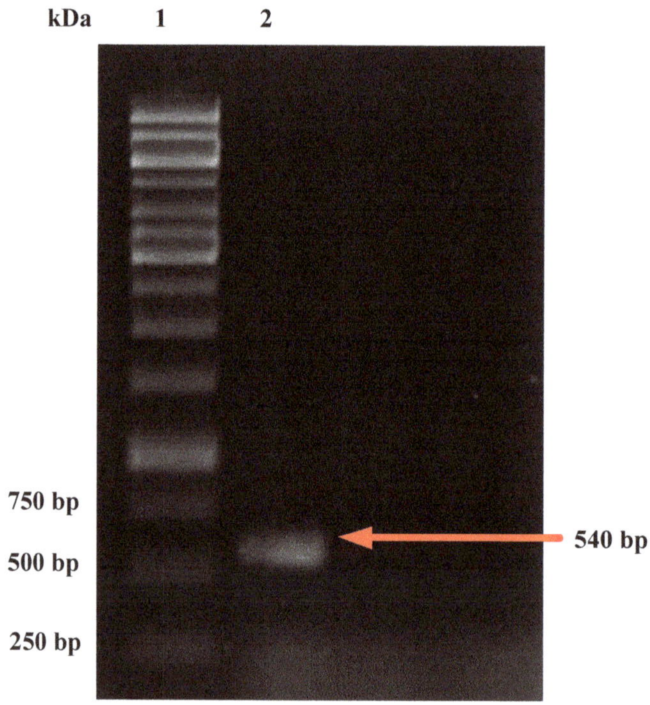

Figure 2. PCR amplification of genomic DNA of *P. citrinum* (lane 2) on agarose gel 1%. Lane 1—DNA size marker.

"CATGCTCCGGCCGCCATGGCGGCCGCGGGAATTCGATTTCCGTAGGTGAACCTGCGGAA GGATCATTACCGAGTGCGGGCCCCTCGGGGCCCAACCTCCCACCCGTGTTGCCCGAACCTATGT TGCCTCGGCGGGCCCCGCGCCCGCCGACGGCCCCCCTGAACGCTGTCTGAAGTTGCAGTCTGA GACCTATAACGAAATTAGTTAAAACTTTCAACAACGGATCTCTTGGTTCCGGCATCGATGAAG AACGCAGCGAAATGCGATAACTAATGTGAATTGCAGAATTCAGTGAATCATCGAGTCTTTGAA CGCACATTGCGCCCTCTGGTATTCCGGAGGGCATGCCTGTCCGAGCGTCATTGCTGCCCTCAAG CCCGGCTTGTGTGTTGGGCCCCGTCCCCCCCGCCGGGGGACGGGCCCGAAAGGCAGCGGCG

GCACCGCGTCCGGTCCTCGAGCGTATGGGGCTTCGTCACCCGCTCTAGTAGGCCCGGCCGGCG CCAGCCGACCCCCAACCTTTAATTATCTCAGGTTGACCT".

To gain insights into the evolutionary relationship, two methods for creation of the phylogenetic trees, namely Neighbor Joining (NJ) and Maximum Parsimony (MP), were employed, resulting in almost identical topologies. The aligned dataset consisted of 42 taxa and 100 characters. Our sequence (marked as unknown) achieved 100% matching with those of *Penicillium citrinum* existing in the database, as evidenced by the excellent bootstrap results. Therefore, the examined fungus presumably belonged to the *Penicillium* genus and firmly aligned with the *P. citrinum* species, as demonstrated by the the strong sequence similarities with the said species.

3.3. Optimization of β-Glucosidase Production by P. citrinum

The optimum incubation time is the time interval at which the highest β-glusosidase activity was attained. After incubation for 24, 48, 72, 96, 120, 144, and 168 h, β-glusosidase activity was 0.23, 2.56, 8.16, 21.25, 23.55, 33.37, and 23.23 U/mL (Figure 3), respectively. This indicates that the highest β-glucosidase activity of 33.63 U/mL was achieved at 144 h of the fermentation time and that prolonged incubation time seemed to reduce enzyme yield. This could be explained by the reduced quantities of nutrients, in both micro and macro forms, existing in the medium after an extended period of fermentation. The phenomenon highlights the role of fungal physiology in inactivating secretary machinery of the enzymes. In addition, the rapid enzyme production in the initial period might be attributable to the high resistance to microbial hydrolysis of the materials, which was later diminished as the fermentation time elapsed.

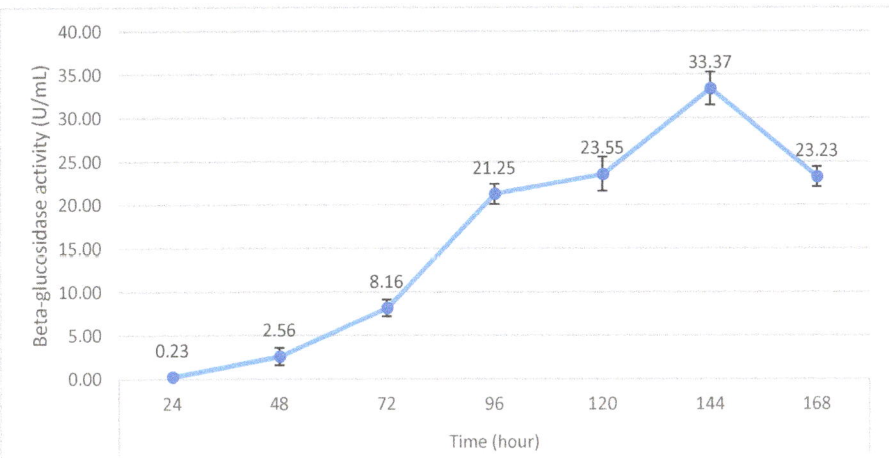

Figure 3. Effect of fermentation time on β-glucosidase activity of *P. citrinum*.

Carbon source is a critical factor affecting the production of enzymes. Therefore, *P. citrinum* was tested to grow in fermentation medium containing various carbon sources under the temperature of 30 °C, agitation speed of 200 rpm, and for six days. Figure 4 shows that the highest β-glucosidase activity of 32.97 U/mL was achieved on potato, and the lower β-glucosidase activities of 4.39, 1.80, and 12.17 U/mL were recorded on barley, corn, and Cz sources, respectively.

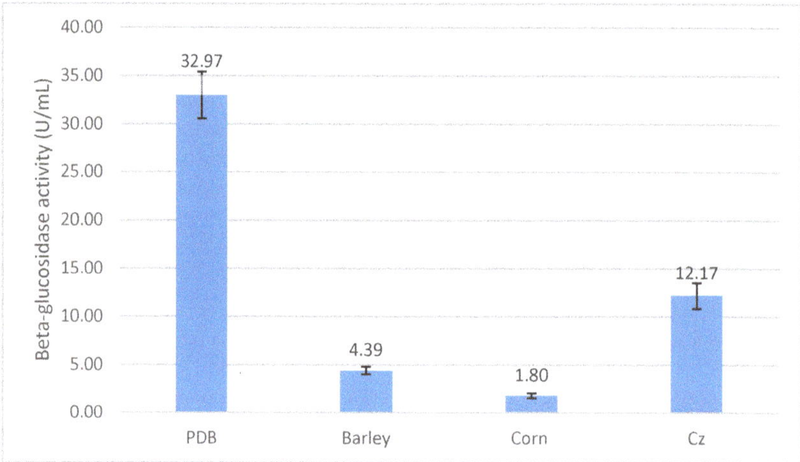

Figure 4. Effect of different carbon sources on β-glucosidase activity of *P. citrinum.*

3.4. Fermentation of Soybean Residual Extract by P. citrinum for Aglycones (Genistein, Daidzein) Production

Figure 5 and Table 1 show that soybean residual extract originally contained glucosides, which were still dominant in the first 24 h fermentation, but partly hydrolyzed after 48 h and completely converted to aglycones after 72 h fermentation. Moreover, the data revealed that at 10–20% (*v*/*v*) volumetric ratio of *P. citrinum* inoculum to CDN substrate, the transformation into aglycones from glucoside forms complete took place during 72 h of fermentation (Table 2). This indicates that the optimal conditions for fermentation of soybean residual extract to aglycones was 10% (*v*/*v*) of *P. citrinum* enzyme to CDN with fermentation time of 72 h at 30 °C.

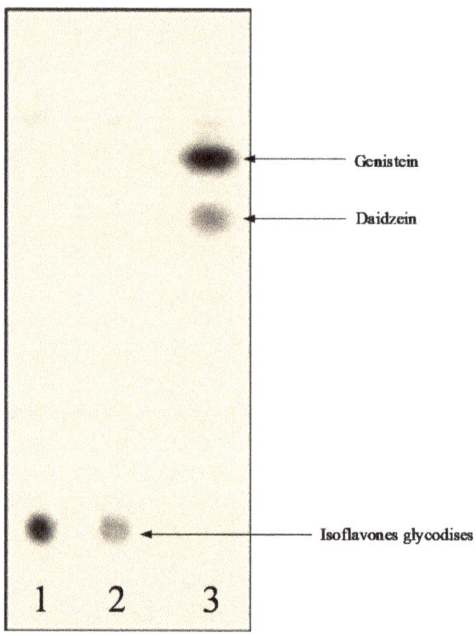

Figure 5. TLC results of fermentation of soybean residual extract using *P. citrinium* at 30 °C during (1) 2 h, (2) 48 h, and (3) 72 h.

Table 1. Effect of hydrolysis time of soybean residual extract on isoflavon glucosides and aglycones formation by *P. citrinium* at 30 °C.

Fermentation Time (h)	Isoflavone Glucosides	Aglycones	Remark
0	++++	-	Original soybean extract without addition of enzyme
24	++++	-	
48	+++	+	
72	-	++++	
96	-	++++	
120	-	++++	

Note: ++++, dark stain; +++, medium stain; +, weak stain; -, no stain.

Table 2. Effect of volumetric ratio of soybean residual extract/enzyme on isoflavone glucosides and aglycones formation by *P. citrinium* at 30 °C.

Soybean Residual Extract/Enzyme Ratio (*v*/*v*, %)	Isoflavone Glucosides	Aglycones	Remark
0	++++	-	Original soybean extract without addition of enzyme
5	-	+	
10	-	+++	
15	+	+++	
20	++	+++	

Note: ++++, dark strain; +++, medium stain; ++, medium-weak stain; +, weak stain; -, no stain.

3.5. Identification and Characterization of Aglycones Using LC-MS/MS, TLC, and NMR

LC-MS/MS results reveal that in the first stage of fragmentation in the negative ion mode, one ion of high intensity at m/z 253 (Figure 6A) was observed, which corresponded to the elemental composition of ($C_{15}H_{10}O_4$, daidzein) with molecular weight of 254.23 (Figure 6B). On the other hand, in the positive ion mode, the highest intensity fragment appeared at m/z 255, corresponding to the same ion having an elemental chemical composition of ($C_{15}H_{10}O_4$). Data indicates that daidzein obtained after extraction and purification from fermentation broth is of high purity. Similarly, genistein ion fragments appeared at m/z 269.0 (Figure 6C) and m/z 271.1 (Figure 6D), having the highest intensity in the negative and positive mode, respectively. These findings are in line with the results of the standard reagents and of previously reported works [3,7].

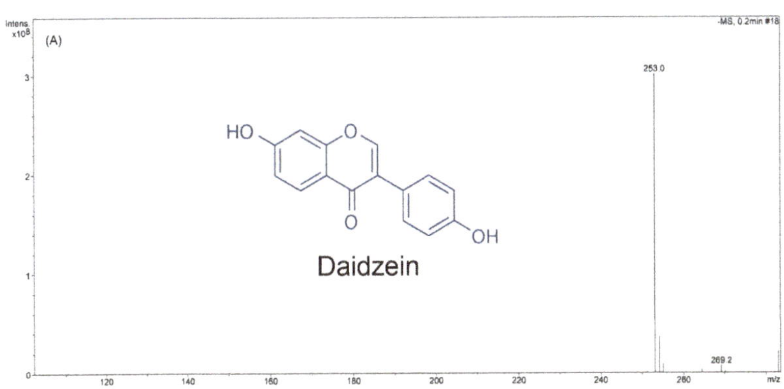

Figure 6. *Cont.*

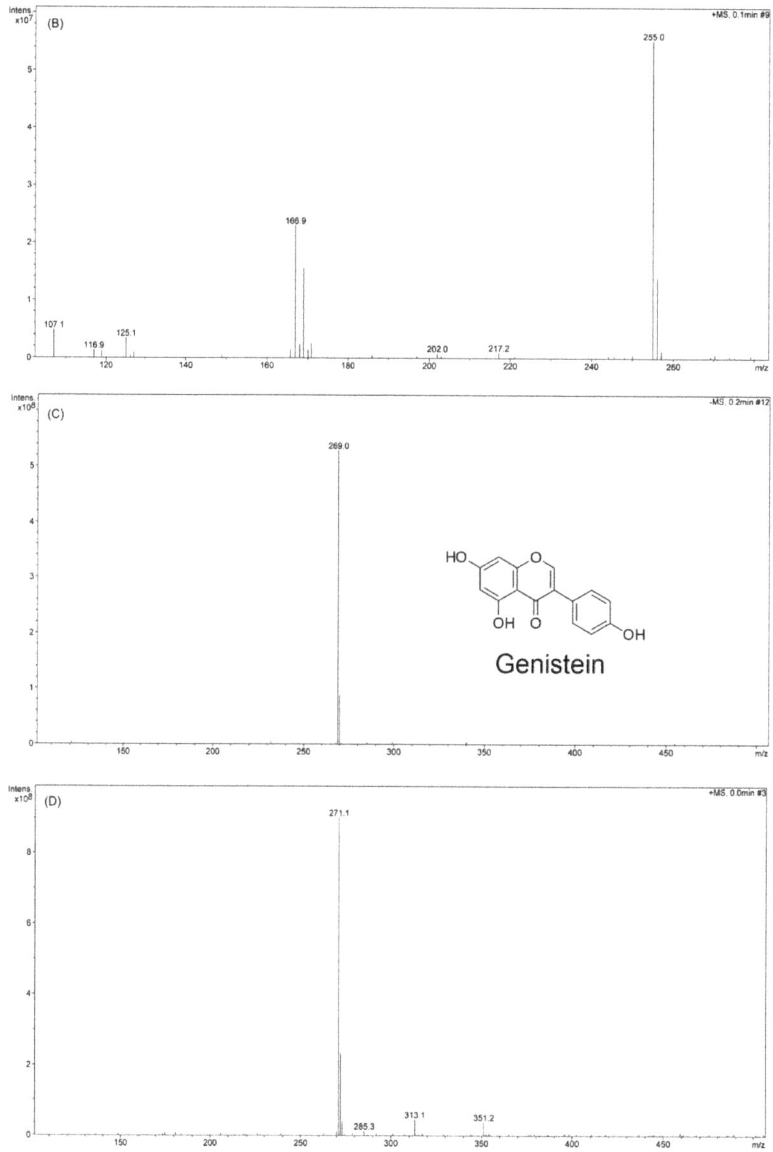

Figure 6. LC-MS/MS spectra of daidzein and genistein. Negative mode of daidzein at m/z 253 (**A**), positive mode of daidzein at m/z 255 (**B**), negative mode of genistein at m/z 269.0 (**C**), and positive mode of genistein at m/z 271.1 (**D**).

For genistein (DTE1, 99.80 mg, purity 90%), light yellow needle crystals, $R_f = 0.37$ (TLC, silica gel, solvent n-hexan/acetone (7/3, *v*/*v*)) appeared as yellow-brown with 5% $FeCl_3$ and dark green with $Ce(SO_4)_2$ (data not shown). ^{1}H-NMR (Acetone, 500 Hz): δ (ppm) 6.28 (1H, d, J = 2 Hz, H6), 6.41 (1H, d, J = 2.5 Hz, H8), 6.90 (2H, dd, J = 2 Hz, H3′, H5′), 7.46 (2H, dd, J = 2 Hz, H2′, H6′), 8.163 (1H, d, J = 1.5 Hz, H2), 13.03 (1H, s, OH). ^{13}C-NMR (Acetone, 125 Hz): δ (ppm) 94.15 (s, C8),

99.52 (s, C6), 105.87 (s, C10), 115.65 (d, C3′, C5′), 122.75 (s, C1′), 123.76 (d, C3), 130.86 (d, C2′, C6′), 153.98 (s, C2), 163.64 (d, C5), 158.76 (d, C9), 158.13 (t, C4), 164.70 (t, C7), 181.35 (t, C7).

For daidzein (DTE2, 300.58 mg, purity 90%); white needle crystals, R_f = 0.30 (TLC, silica gel, solvent n-hexan/acetone (7/3, *v*/*v*)) appeared as dark-green with $Ce(SO_4)_2$ and colorless with 5% $FeCl_3$ (data not shown). ^{1}H-NMR (Acetone, 500 Hz): δ (ppm) 6.89 (3H, dd, J = 2 Hz, 9 Hz, H3′, H5′, OH4′), 6.98 (1H, m, H8), 7.47 (2H, dd, J = 5.5 Hz, 2 Hz, H2′, H6′), 8.06 (1H, d, J = 8.5 Hz, H6), 8.14 (1H, s, H2). ^{13}C-NMR (DMSO, 125 Hz): δ (ppm) 102.04 (s, C8), 115.07 (d, C6), 116.58 (d, C5′, C3′), 123.43 (d, C3), 127.23 (t, C1′), 129.99 (d, C5), 125.74 (s, C2), 157.11 (d, C4′), 157.37 (d, C9), 162.45 (t, C7), 174.63 (s, C4).

4. Discussion

It was reported that β-glucosidase can be produced on different carbon sources by fungi with different activities [36]. Lachke et al. [37] found that the maximum β-glucosidase activity of *Penicillium funiculosum* was 30–36 U/mL, achieved on 3% rice bran or defatted oil cakes after 288 h fermentation. By stark contrast, the peak activity was found at a much lower level of 2.8 U/mL with *Penicillium miczynskii* cultured on 3% pineapple peel within 216 h of fermentation (Table 3) [38]. Furthermore, the investigation of Jeya et al. [39] resulted in the highest β-glucosidase activity of 26.4 U/mL in *P. purpurogenum* produced on rice straw, while Ng et al. [40] achieved the highest β-glucosidase activity of 57.5 U/g solid when rice bran was used via solid fermentation. Although other variables (e.g., temperature, pH, etc.) have not been optimized in this study, β-glucosidase produced by the isolated *P. citrinum* was reached 33.63–33.72 U/mL, listing *P. citrinum* as the top fungal strain among the reported fungi producing highest activity of β-glucosidase. In terms of scalability, β-glucosidase holds the potential to be cloned into other hosts such as *Escherichia coli* for large-scale industrial production [41].

Table 3. β-glucosidase production of several fungal strains.

Fungal Strain	Fermentation Conditions	β-Glucosidase Activity (U/mL)	Ref.
P. funiculosum	Substrate: Rice bran, defatted oil cakes Temperature: 20 °C pH: 4.5 Fermentation time: 288 h	30–36	[34]
P. miczynskii	Substrate: Pineapple peel Temperature: 30 °C pH: 5.5 Fermentation time: 216 h	2.82	[35]
P. purpurogenum KJS506	Substrate: Rice straw Temperature: 32 °C pH: 4 Fermentation time: 144 h	26.4	[36]
P. citrinium YS40-5	Substrate: Rice bran Temperature: 70 °C pH: 6.0 Fermentation time: 96 h	57.5 U/g	[37]
P. citrinium	Substrate: PDB Temperature: 30 °C pH: 4.5 Fermentation time: 144 h	33.72	Current study

The hydrolysis yield of soybean waste extract (rich in isoflavone glycosides) to their aglycones by *P. citrinum* was estimated to be 98.7%, which is comparable to aglycone rates of 94.22 and 97.14% achieved by *D. hansenii* UFV-1 immobilized cells containing β-glycosidase and free enzyme, respectively [42], on soy molasses. However, our reported hydrolysis rate is higher than the rate of 93% achieved in deglycosylation of extracts of soybean flour and embryo in β-glucosidase derived from *Paecilomyces thermophila* J18 [43]. Moreover, the biotransformation rate is also higher than that by a marine *Streptomyces* sp. 060524, which achieved only 90% on the same substrate during 108 h fermentation [5].

Isoflavone aglycones can be derived from soybean waste by extraction with ethanol/water solvent followed by acid hydrolysis and purification [44], reaching purity of 92%. Alternatively, the soybean extract can be hydrolyzed by cellulase to produce crude aglycone isoflavones, followed by purification, to obtain purity of 80.38–87.21%. Experimental data obtained in this study showed that the purity of genistein and daidzein reached over 90%, which is comparable to the purity obtained by the reported methods, demonstrating raw materials with potential for utilization in industrial applications, e.g., pharmaceutics and functional foods, with minimum refining [3].

5. Conclusions

The *P. citrinum* fungus strain that produces the highest β-glucosidase activity of 33.72 U/mL, thus placing it amongst the most active fungi in this regard, has been successfully isolated from roots of *Clerodendron cyrtophyllum* Turcz. The fungus demonstrated a catalytic capacity to hydrolyze isoflavones-rich soybean extract into aglycones (e.g., genistein and daidzein) with a hydrolysis yield of 98.7% after 72 h of fermentation at 30 °C. Purification of the hydrolyzed mixture (rich in genistein and daidzein) by column chromatography using hexane/acetone (29:1/1:1) resulted in aglycone products with purity of over 90%. These results imply that soybean extract is a promising raw material for manufacture of functional foods derived from aglycones. Our study demonstrated a potential pathway for the production of aglycones from residual soybean extract via fermentation with an isolated fungi as a biocatalyst, with applciations in the pharmaceutical and functional food industries.

Author Contributions: Investigation, D.T.D., D.P.L., T.D.N., B.H.T., H.M.P.T., H.N.D., V.H.L., T.D.P., V.T.T., M.Q.P. and H.H.P.T.; Supervision, Q.T.T.; Writing–Original draft, D.T.D.

Funding: This work was financially supported by Vietnam Academy of Science and Technology under the grant No. VAST.ĐLT.14/14-15.

Conflicts of Interest: The authors declare no conflict of interest.

References

1. Anderson, J.J.B.; Garner, S.C. Phytoestrogens and human function. *Nutr. Today* **1997**, *32*, 232. [CrossRef]
2. Markovic, R.; Baltic, M.Z.; Pavlovic, M.; Glisic, M.; Radulovic, S.; Djordjevic, V.; Sefer, D. Isoflavones-from biotechnology to functional foods. *Procedia Food Sci.* **2015**, *5*, 176–179. [CrossRef]
3. Yatsu, F.K.J.; Koester, L.S.; Bassani, V.L. Isoflavone-aglycone fraction from Glycine max: A promising raw material for isoflavone-based pharmaceutical or nutraceutical products. *Rev. Bras. Farmacogn.* **2016**, *26*, 259–267. [CrossRef]
4. Hong, J.L.; Qin, X.Y.; Shu, P.; Wang, Q.; Zhou, Z.F.; Wang, G.K.; Lin, B.B.; Wang, Q.; Qin, M.J. Comparative study of isoflavones in wild and cultivated soybeans as well as bean products by high-performance liquid chromatography coupled with mass spectrometry and chemometric techniques. *Eur. Food Res. Technol.* **2011**, *233*, 869. [CrossRef]
5. Hu, S.C.; Hong, K.; Song, Y.C.; Liu, J.Y.; Tan, R.X. Biotransformation of soybean isoflavones by a marine Streptomyces sp. 060524 and cytotoxicity of the products. *World J. Microbiol. Biotechnol.* **2008**, *25*, 115. [CrossRef]
6. Baú, T.R.; Ida, E.I. Soymilk processing with higher isoflavone aglycone content. *Food Chem.* **2015**, *183*, 161–168. [CrossRef] [PubMed]
7. Delmonte, P.; Perry, J.; Rader, J.I. Determination of isoflavones in dietary supplements containing soy, red clover and kudzu: Extraction followed by basic or acid hydrolysis. *J. Chromatogr. A* **2006**, *1107*, 59–69. [CrossRef] [PubMed]
8. Rostagno, M.A.; Villares, A.; Guillamón, E.; García-Lafuente, A.; Martínez, J.A. Sample preparation for the analysis of isoflavones from soybeans and soy foods. *J. Chromatogr. A* **2009**, *1216*, 2–29. [CrossRef] [PubMed]

9. Handa, C.L.; Couto, U.R.; Vicensoti, A.H.; Georgetti, S.R.; Ida, E.I. Optimisation of soy flour fermentation parameters to produce β-glucosidase for bioconversion into aglycones. *Food Chem.* **2014**, *152*, 56–65. [CrossRef]
10. Chun, J.; Kim, G.M.; Lee, K.W.; Choi, I.D.; Kwon, G.H.; Park, J.Y.; Jeong, S.J.; Kim, J.S.; Kim, J.H. Conversion of isoflavone glucosides to aglycones in soymilk by fermentation with lactic acid bacteria. *J. Food Sci.* **2007**, *72*, M39–M44. [CrossRef]
11. Aguieiras, E.C.G.; Cavalcanti-Oliveira, E.D.; de Castro, A.M.; Langone, M.A.P.; Freire, D.M.G. Simultaneous enzymatic transesterification and esterification of an acid oil using fermented solid as biocatalyst. *J. Am. Oil Chem. Soc.* **2017**, *94*, 551–558. [CrossRef]
12. Papadaki, A.; Kopsahelis, N.; Mallouchos, A.; Mandala, I.; Koutinas, A.A. Bioprocess development for the production of novel oleogels from soybean and microbial oils. *Food Res. Int.* **2019**, *126*, 108684. [CrossRef]
13. Papadaki, A.; Androutsopoulos, N.; Patsalou, M.; Koutinas, M.; Kopsahelis, N.; Castro, A.; Papanikolaou, S.; Koutinas, A. Biotechnological production of fumaric acid: The effect of morphology of rhizopus arrhizus NRRL 2582. *Fermentation* **2017**, *3*, 33. [CrossRef]
14. Papadaki, A.; Papapostolou, H.; Alexandri, M.; Kopsahelis, N.; Papanikolaou, S.; de Castro, A.M.; Freire, D.M.G.; Koutinas, A.A. Fumaric acid production using renewable resources from biodiesel and cane sugar production processes. *Environ. Sci. Pollut. Res.* **2018**, *25*, 35960–35970. [CrossRef]
15. Haas, M.J. Improving the economics of biodiesel production through the use of low value lipids as feedstocks: Vegetable oil soapstock. *Fuel Process. Technol.* **2005**, *86*, 1087–1096. [CrossRef]
16. Aguieiras, E.C.G.; Papadaki, A.; Mallouchos, A.; Mandala, I.; Sousa, H.; Freire, D.M.G.; Koutinas, A.A. Enzymatic synthesis of bio-based wax esters from palm and soybean fatty acids using crude lipases produced on agricultural residues. *Ind. Crops Prod.* **2019**, *139*, 111499. [CrossRef]
17. Fujita, A.; Alencar, S.M.; Park, Y.K. Conversion of isoflavone glucosides to aglycones by partially purified β-glucosidases from microbial and vegetable sources. *Appl. Biochem. Biotechnol.* **2015**, *176*, 1659–1672. [CrossRef]
18. Ahn-Jarvis, J.H.; Teegarden, M.D.; Schwartz, S.J.; Lee, K.; Vodovotz, Y. Modulating conversion of isoflavone glycosides to aglycones using crude beta-glycosidase extracts from almonds and processed soy. *Food Chem.* **2017**, *237*, 685–692. [CrossRef]
19. da Silva, L.H.; Celeghini, R.M.S.; Chang, Y.K. Effect of the fermentation of whole soybean flour on the conversion of isoflavones from glycosides to aglycones. *Food Chem.* **2011**, *128*, 640–644. [CrossRef]
20. Xie, L.; Hettiarachchy, N.S.; Cai, R.; Tsuruhami, K.; Koikeda, S. Conversion of Isoflavone Glycosides to Aglycones in SoyLife and Soymeal Using β-glycosidase. *J. Food Sci.* **2003**, *68*, 427–430. [CrossRef]
21. Chang, F.; Xue, S.; Xie, X.; Fang, W.; Fang, Z.; Xiao, Y. Carbohydrate-binding module assisted purification and immobilization of β-glucosidase onto cellulose and application in hydrolysis of soybean isoflavone glycosides. *J. Biosci. Bioeng.* **2018**, *125*, 185–191. [CrossRef] [PubMed]
22. Molina, G.; de Lima, E.A.; Borin, G.P.; de Barcelos, M.C.S.; Pastore, G.M. Chapter 7—Beta-Glucosidase From Penicillium. In *New and Future Developments in Microbial Biotechnology and Bioengineering*; Gupta, V.K., Rodriguez-Couto, S., Eds.; Elsevier: Amsterdam, The Netherlands, 2018; pp. 137–151.
23. Yan, F.; Xia, W.; Zhang, X.; Chen, S.; Nie, X.; Qian, L. Characterization of β-glucosidase from Aspergillus terreus and its application in the hydrolysis of soybean isoflavones. *J. Zhejiang Univ. Sci. B* **2016**, *17*, 455–464. [CrossRef] [PubMed]
24. Chang, K.H.; Jo, M.N.; Kim, K.T.; Paik, H.D. Evaluation of glucosidases of Aspergillus niger strain comparing with other glucosidases in transformation of ginsenoside Rb1 to ginsenosides Rg3. *J. Ginseng Res.* **2014**, *38*, 47–51. [CrossRef] [PubMed]
25. Baraldo Junior, A.; Borges, D.G.; Tardioli, P.W.; Farinas, C.S. Characterization of β-Glucosidase Produced by Aspergillus niger under Solid-State Fermentation and Partially Purified Using MANAE-Agarose. *Biotechnol. Res. Int.* **2014**, *2014*, 317092. [CrossRef] [PubMed]
26. Langston, J.; Sheehy, N.; Xu, F. Substrate specificity of Aspergillus oryzae family 3 β-glucosidase. *Biochim. Biophys. Act. BBA Proteins Proteom.* **2006**, *1764*, 972–978. [CrossRef]
27. Krogh, K.B.R.M.; Harris, P.V.; Olsen, C.L.; Johansen, K.S.; Hojer-Pedersen, J.; Borjesson, J.; Olsson, L. Characterization and kinetic analysis of a thermostable GH3 β-glucosidase from Penicillium brasilianum. *Appl. Microbiol. Biotechnol.* **2010**, *86*, 143–154. [CrossRef] [PubMed]

28. Tsukada, T.; Igarashi, K.; Yoshida, M.; Samejima, M. Molecular cloning and characterization of two intracellular β-glucosidases belonging to glycoside hydrolase family 1 from the basidiomycete Phanerochaete chrysosporium. *Appl. Microbiol. Biotechnol.* **2006**, *73*, 807. [CrossRef]
29. Kudo, K.; Watanabe, A.; Ujiie, S.; Shintani, T.; Gomi, K. Purification and enzymatic characterization of secretory glycoside hydrolase family 3 (GH3) aryl β-glucosidases screened from Aspergillus oryzae genome. *J. Biosci. Bioeng.* **2015**, *120*, 614–623. [CrossRef]
30. Li, D.; Li, X.; Dang, W.; Tran, P.L.; Park, S.H.; Oh, B.C.; Hong, W.S.; Lee, J.S.; Park, K.H. Characterization and application of an acidophilic and thermostable β-glucosidase from Thermofilum pendens. *J. Biosci. Bioeng.* **2013**, *115*, 490–496. [CrossRef]
31. Dekker, R.F.H. Kinetic, inhibition, and stability properties of a commercial β-D-glucosidase (cellobiase) preparation from *aspergillus niger* and its suitability in the hydrolysis of lignocellulose. *Biotechnol. Bioeng.* **1986**, *28*, 1438–1442. [CrossRef]
32. Tendulkar, S.R.; Saikumari, Y.K.; Patel, V.; Raghotama, S.; Munshi, T.K.; Balaram, P.; Chattoo, B.B. Isolation, purification and characterization of an antifungal molecule produced by Bacillus licheniformis BC98, and its effect on phytopathogen Magnaporthe grisea. *J. Appl. Microbiol.* **2007**, *103*, 2331–2339. [CrossRef] [PubMed]
33. Kidd, S.; Halliday, C.L.; Alexiou, H.; Ellis, D.H. *Descriptions of Medical Fungi*; CutCut Digital: Richmond, VA, USA, 2016.
34. White, T.J.; Bruns, T.; Lee, S.J.W.T.; Taylor, J. Amplification and direct sequencing of fungal ribosomal RNA genes for phylogenetics. In *PCR Protocols: A Guide to Methods and Applications*; Innis, M.A., Gelfand, D.H., Sninsky, J.J., White, T.J., Eds.; Academic Press: Cambridge, MA, USA, 1990; pp. 315–322.
35. Cai, Y.J.; Buswell, J.A.; Chang, S.T. β-Glucosidase components of the cellulolytic system of the edible straw mushroom, Volvariella volvacea. *Enzyme Microb. Technol.* **1998**, *22*, 122–129. [CrossRef]
36. So, J.H.; Kim, W.C.; Shin, J.H.; Yu, C.B.; Rhee, I.K. Purification and characterization of a β-glucosidase capable of hydrolyzing soybean isoflavone glycosides from Pichia guilliermondii K123-1. *Food Sci. Biotechnol.* **2010**, *19*, 1373–1379. [CrossRef]
37. Lachke, A.H.; Bastawde, K.B.; Powar, V.K.; Srinivasan, M.C. Enhanced production of extracellular β-glucosidase by penicillium funiculosum in submerged culture. *Biotechnol. Lett.* **1983**, *5*, 649–652. [CrossRef]
38. Beitel, S.M.; Knob, A. Penicillium miczynskii β -glucosidase: A Glucose-Tolerant Enzyme Produced Using Pineapple Peel as Substrate. *Ind. Biotechnol.* **2013**, *9*, 293–300. [CrossRef]
39. Jeya, M.; Joo, A.R.; Lee, K.M.; Tiwari, M.K.; Lee, K.M.; Kim, S.H.; Lee, J.K. Characterization of β-glucosidase from a strain of Penicillium purpurogenum KJS506. *Appl. Microbiol. Biotechnol.* **2010**, *86*, 1473–1484. [CrossRef] [PubMed]
40. Ng, I.S.; Li, C.W.; Chan, S.P.; Chir, J.L.; Chen, P.T.; Tong, C.G.; Yu, S.M.; Ho, T.H.D. High-level production of a thermoacidophilic β-glucosidase from Penicillium citrinum YS40-5 by solid-state fermentation with rice bran. *Bioresour. Technol.* **2010**, *101*, 1310–1317. [CrossRef]
41. Xue, Y.; Song, X.; Yu, J. Overexpression of β-glucosidase from Thermotoga maritima for the production of highly purified aglycone isoflavones from soy flour. *World J. Microbiol. Biotechnol.* **2009**, *25*, 2165–2172. [CrossRef]
42. Maitan-Alfenas, G.P.; de A Lage, L.G.; de Almeida, M.N.; Visser, E.M.; de Rezende, S.T.; Guimarães, V.M. Hydrolysis of soybean isoflavones by Debaryomyces hansenii UFV-1 immobilised cells and free β-glucosidase. *Food Chem.* **2014**, *146*, 429–436. [CrossRef]
43. Yang, S.; Wang, L.; Yan, Q.; Jiang, Z.; Li, L. Hydrolysis of soybean isoflavone glycosides by a thermostable β-glucosidase from Paecilomyces thermophila. *Food Chem.* **2009**, *115*, 1247–1252. [CrossRef]
44. Nemitz, M.C.; Teixeira, H.F.; von Poser, G.L. A new approach for the purification of soybean acid extract: Simultaneous production of an isoflavone aglycone-rich fraction and a furfural derivative-rich by-product. *Ind. Crops Prod.* **2015**, *67*, 414–421. [CrossRef]

foods MDPI

Article

Improved Lycopene Production from Different Substrates by Mated Fermentation of *Blakeslea Trispora*

Ayse Sevgili and Osman Erkmen *

Department of Food Engineering, Faculty of Engineering, University of Gaziantep, 27310 Gaziantep, Turkey; aysesevgili8@gmail.com
* Correspondence: erkmen@gantep.edu.tr; Tel.: +90-342-317-2313; Fax: +90-342-317-2362

Received: 23 February 2019; Accepted: 28 March 2019; Published: 9 April 2019

Abstract: The production of lycopene from different substrates by *Blakeslea trispora* in fermentation was investigated. Lycopene productions from 4 and 6% glucose (pH 6.5) in shake flask fermentation were 77.7 and 28.1 mg L^{-1}. Increasing the glucose concentration to 6% resulted in a decrease in lycopene production by 36.2%. A maximum lycopene concentration of 944.8 mg L^{-1} was detected with 4% glucose supplemented with 1.0 % sunflower oil in fermentor studies. Lycopene productions in the presence of sunflower and corn oils in the fermentor were 12.2 and 11.1 times higher, respectively, then without oil from 4 % glucose in a shake flask. Lycopene production from orange peel was two times higher in the fermentor than in the shake flask. Zygospores of *B. trispora* are the morphological forms, which are responsible for the production of the lycopene. The highest level of zygospores was correlated with the highest amount of intracellular lycopene in the total biomass dry weight. The media containing only orange powder (1%) gave a 4.9 mg L^{-1} lycopene production in a fermentor. The biosynthesis of lycopene has been started in most cases simultaneously in the early growth phase even in trace amounts. Maximum lycopene concentration was obtained when the medium was supplied with sunflower and corn oils. There is an indirect relationship between biomass and lycopene concentration.

Keywords: lycopene; *Blakeslea trispora*; fermentation; vegetable oil; orange powder

1. Introduction

Lycopene is an important naturally occurring intermediate metabolite in the mevalonate metabolic pathway of microorganisms and can be transformed into β-carotene by lycopene cyclase. Thus, in order to enhance lycopene accumulation, lycopene cyclase activity should be inhibited by adding lycopene cyclase inhibitors into the fermentation medium. Lycopene has important physiological functions to prevent chronic diseases, including osteoporosis, certain types of cancer, cataract formation and cardiovascular diseases; enhance immune responses; and other biological effects in humans and animals [1,2]. In the last few years, demand on lycopene has increased due to its use in the food, pharmaceutical, cosmetic and animal feed industries [3]. Moreover, lycopene is used as an antioxidant to reduce cellular or tissue damage, and as a coloring agent for food products [4]. These make lycopene very attractive in the food and feed industries, and in medicine and cosmetic formulation. Lycopene used industrially is manufactured from plant extraction and chemical synthesis. Market needs for lycopene are not met using plant extraction due to the restricted number of plant sources and high cost. Chemical synthesis also has disadvantages including low yields, product instability, low product quality and high production cost. Due to decreased natural production of lycopene and its increased worldwide demand, there is the potential for lycopene to be produced using fermentation. Various algae and fungi produce intracellular carotenoids, including β-carotene, lycopene and astaxanthin.

Blakeslea trispora, *Mucor circinelloides*, *Candida utilis* and *Phycomyces blakesleeanus* have the ability to produce lycopene [5]. *B. trispora* has an advantage in that it does not need any specific fermentation conditions for growth. Mating type of (+) and (−) *B. trispora* strains together can increase production of lycopene [6].

There is much interest in biotechnological production of lycopene due to consumer demand for high quality and "natural" food additives. Efforts have been made to increase the lycopene yield and reduce process costs, by using low-cost agro-food media rich in organic compounds (such as carbon, nitrogen, mineral and other sources). The amount of lycopene production depends on microorganism, the culture medium, substrates and the fermentation conditions (temperature, pH, aeration rate etc.). Most of the research has been aimed at optimizing the culture conditions that directly affect the growth of the microorganism. The main research is focused on optimization of the medium, extraction of lycopene from *B. trispora* and fermentation conditions [4–9].

An industrial mold over-producing lycopene is *B. trispora* which accumulates all-trans-carotene, as a secondary metabolite [10]. *B. trispora* is a saprophyte and performs vegetative cycle of spores and filamentous mycelia [6]. There has been a lot of effort to use food wastes in the production of lycopene mitigation [11]. At present, there are no research on the use of orange powder and sunflower oil in the production of lycopene. The transport cost, sales problems, low quality and susceptibility to spoilage of the food-wastes by microorganisms have led to alternative utilization approaches. Agro-food wastes (fruits and vegetables: cabbage residues, watermelon husk and peach powders) have been used in the production of carotenes by *B. trispora* [12,13].

Optimization of fermentation conditions can also be required in the production of lycopene from different nutrient sources. This report was focused on the effect of a series of media containing various substrates (glucose, refined vegetable oils and orange powder) in the accumulation of lycopene in submerged shake flasks and fermentor studies. Morphological, biomass and pH changes occurring during fermentations were also studied.

2. Materials and Methods

2.1. Microorganism and Chemicals

Blakeslea trispora mating strains ATCC 14,271 (+) and ATCC 14,272 (−) were obtained from the American Type Culture Collection (ATCC; Rockville, MA, USA). Lyophilized cultures were rehydrated with malt extract broth (MEB; Darmstadt, Germany) by incubation at 35 °C for 3 days and were used in the stock culture preparation. Stock cultures of *B. trispora* were prepared on potato dextrose agar (PDA; Darmstadt, Germany) slant. Yeast extract, glucose, agar, asparagine, starch, lycopene, nicotine, $KHPO_4$, $MgSO_4$, HPLC grade petroleum ether, methanol, acetonitrile and dichloromethane were obtained from Sigma-Aldrich (Interlab Company, Adana, Turkey).

2.2. Preparation of B. Trispora Cultures

B. trispora (+) and (−) cultures were prepared for fermentation studies using a yeast phosphate soluble starch (YpSS) agar medium. The composition of YpSS agar medium (g L^{-1}) was: Yeast extract 4.0, K_2HPO_4 1.0, $MgSO_4$ 0.5, starch 15.0 and agar 20.0. The pH of the medium was adjusted to 6.5. Each of the *B. trispora* strains from stock culture was inoculated separately onto YpSS agar plates and incubated at 28 °C for 5 days. After sporulation of cultures, 10 mL of sterile distilled water was added onto the cultures in the Petri plates and the spores were collected by scraping off the medium surface using a sterile cotton swab. The spore suspension containing 3.9–8.5 $\times$ 10^6 spores ml^{-1} was used to inoculate the fermentation media. Strains were inoculated into the fermentation medium with 2.5 and 7.5 mL of spore suspension (+) and (−) strains, respectively, to provide 1:3 ratio of inoculation.

2.3. Preparation of Fermentation Media

The basal fermentation medium (BFM) had the following composition (g L^{-1}): Yeast extract 1.0, asparagine 2.0, K_2HPO_4 1.5 and $MgSO_4$ 0.5. BFM was supplemented with different concentrations of ingredients as given in Table 1: Refined vegetable oils (sunflower and corn oils), glucose and milled orange peel (powder). Samples of orange (*Citrus sinensis*) peel were dried at 40 °C to reach a final moisture lower than 10%, milled to a particle size (250 μm) using a laboratory grinder (Roller type lab mill RM 1300, Erkaya, Ataşehir, İstanbul, Turkey), homogenized in a single lot to avoid any variation in composition, and stored at 4 °C in a cold chamber until use. Fermentation media were prepared at two different pH: 6.5 and 5.5. Fermentation media (50 mL) in Erlenmeyer flasks (250 mL) were sterilized at 121 °C for 15 min in autoclave. Fermentation media for fermentor studies were sterilized in a 7 L Bioflow 410 sterilizable-in-place benchtop fermentor (New Brunswick Scientific BioFlo 410 SIP Fermentor, Eppendorf, New York, MA, USA).

Table 1. Medium composition.

Medium Number	Composition			
	Glucose (g L^{-1})	Sunflower Oil (ml L^{-1})	Corn Oil (ml L^{-1})	Orange Powder (g L^{-1})
		In shake flask studies		
1	40	-	-	-
2	60	-	-	-
3	40	10	-	-
4	40	30	-	-
6	40	-	10	-
7	40	-	30	-
8	-	-	-	10
9	-	-	-	20
		In fermentor studies		
10	40 [a,b]	-	-	-
11	60 [a]	-	-	-
12	40 [a]	10	-	-
13	40 [a]	-	10	-
14	-	-	-	10 [a]

[a] At 3 L min^{-1} aeration and 500 rpm agitation. [b] At 12 L min^{-1} aeration and 750 rpm agitation.

2.4. Shake Flask Fermentation Process

Fermentation media (50 mL) in Erlenmeyer flasks (250 mL) were inoculated with 10 mL *B. trispora* strains ATCC 14,271 (+) and 14,272 (−) by 1:3 ratio from spore stock culture. The lycopene cyclase inhibitor of nicotine (5 mM) was added on the second day of fermentation. Fermentation media (total volume 60 mL) were incubated at 28 °C for 7 days on a bench type water bath shaker ST-402 (NÜVE; Sanayi Malzemeleri Imalat ve Ticaret A. Ş., İstanbul, Turkey) with wrist shaking at 120 rpm.

2.5. Fermentation Process in Fermentor

All fermentations were performed in the Bioflow 410 fermentor with a working volume of 5.0 L. Sterilization sequences in this fermentor are fully automatic and heat-up and cool down in under 30 min. All easily initiated using the advanced touch-screen controller. The fermenter was equipped with a pH probe, dissolved oxygen probes, automatic sampler, temperature probe, inoculation port, sparger, exhaust condenser, agitation, vessel light, level sensor and BioCommand SCADA Software (RS-232; New Brunswick Scientific BioFlow 410 SIP Fermentor, Eppendorf, New York, MA, USA).

Four liter of fermentation media (from BFM) were prepared in the fermentor. The initial pH of fermentation was adjusted to 6.5. The fermentation with 4% glucose media was also performed at two aerations and agitations. The fermentor was incubated at 28 °C. The inoculation ratio of two strains (+/−) into the fermentation medium from plate cultures was the same as those in the shake flask. The lycopene cyclase inhibitor of nicotine (5 mM) was added on the second day of fermentation. Dissolved oxygen tension was not controlled during fermentation.

2.6. Sampling

Fermented samples were aseptically removed from the shake flask and fermentor through automatic sampler after 2, 3, 4, 6 and 7 days of fermentation. Initial samples were also removed at the beginning of fermentation. At every sampling interval, 5 mL of sample was taken from the fermentation media under aseptic conditions and the contents were analyzed for lycopene, pH and biomass changes. Two ml of the sample was used in lycopene analysis and 3 mL was used in pH and biomass analysis. At each sampling time and after the first day, 0.4 mL of the sample was also removed for morphological analysis.

Two samples were removed and two parallel analyses were performed from each sample. All experiments were repeated three times.

2.7. Analysis

2.7.1. Lycopene Analysis

Samples were analyzed for lycopene by a High-Performance Liquid Chromatography (HPLC) method [1]. Two ml of a 5 mL sample was mixed with 15 mL of petroleum ether. The mixture was subjected to ultrasound (Soniprep 150 Ultrasonic Disintegrator, MSE, London, UK) for 30 s and then, the sample was centrifuged using a table-type centrifuge (Hettich eba III) at 6000 rpm for 15 min. The supernatant was passed through 4 μm filter paper. About 2.5 mL of liquid was added into a 3 mL vial and the sample was analyzed by HPLC immediately. The presence of lycopene was detected by HPLC using a fluorescence detector. The flow rate was 1 mL min^{-1} and the column was nukleosil C18 (250 × 4.6 mm ID). Temperature was 28 °C. Lycopene was detected at 450 nm and 20 μL sample was injected to HPLC automatically. The result was read for 40 min. The mobile phase was acetonitrile-methanol-water-dichloromethane (7:1.5:0.5:1, v:v:v:v). The mobile phase was filtered through a disposable filter unit (0.45 μm) and degassed in a degasser (Bransonic 2200, 41 Eagle Road Danbury, Connecticut, 06810-1961, CT, USA). The peaks in a chromatogram were evaluated according to the lycopene standard curve.

Analytical Method Validation: Linearity was determined between 0.1 to 0.0001 g ml^{-1} using five levels of calibration in triplicate. The Mandel's fitting test was used to evaluate the linearity of the straight-line regression model [14].

Limit of detection (LOD) and limit of quantification (LOQ) were determined according to the method described by Cucu et al. [15]. For this, standards of 0.1 to 0.0001 g lycopene ml^{-1} were prepared and injected three times each. The mean of the slopes (S) and standard deviation of the intercepts (σ) were calculated from the three calibration curves. The LOD and the LOQ were calculated according to the formulas below [15]:

$$LOD = (3.3 * \sigma)/S$$
$$LOQ = (10 * \sigma)/S$$

For the assessment of the matrix effect, calibration curves were prepared in the petroleum ether and analyzed in triplicates. The calibration curve of lycopene was the equations: y = ((area+intercept)/slope).

2.7.2. pH, Biomass and Morphological Analysis

In biomass analysis, Whatman No. 41 filter paper was dried at 105 °C to constant weight, cooled in a desiccator and weighed before use. Three ml of the remaining sample was filtered through the filter paper. The filter cake on paper was washed three times with distilled water. The filter paper was placed into an incubator at 105 °C, dried until constant weight and then weighed to calculate the biomass (g L^{-1}). The medium with %10 orange powder was used as a control; in biomass detection, this was subtracted from biomass.

In pH analysis, a filtered solution was filled to 10 mL with distilled water and pH was detected using a pH meter (EMAF EM78X model, Interlab Company, Adana, Turkey) equipped with a glass electrode.

In image analysis, the morphological changes of *B. trispora* during the fermentation was analyzed using an image analysis system consisting of a light microscope (Olympus BX51; Olympus Corporation, Shinjuku Monolith, 3-1, Nishi Shinjuku 2-chome, Shinjuku-ku, Tokyo, Japan) equipped with a Pixera PVC 100C camera. Samples for image analysis were prepared by mixing 0.4 mL of fermentation broth with 0.1 mL of lactophenol blue. One drop of the mixed sample was pipetted on to a microscope slide, covered by a cover slip so that no air gap or bubbles were trapped between the sample and the coverslip and observed immediately with 20X objective by taking 5 different images. The morphological parameters measured were the area of zygospores and zygospores (% of total area of mycelium), the diameter of zygospores (mm), freely dispersed hyphae and clumps (aggregates) of different sizes on the basis of their projected area, and the hyphae length (mm). The microscope eyepiece is equipped with a scale micrometer that is used to measure the morphological parameters of mold. For each sample, the process was repeated at least 5 times using new positions on the same slide, and the morphological parameters were expressed as the mean values of each sample [16].

2.8. Statistical Analysis

All data were analyzed by SPSS 16 software (SPSS Inc., Chicago, IL, USA). Statistical differences between fermentation time (days) and between substrates used in the fermentations were tested by one-way analysis of variance following Duncan's multiple-range test. A probability of 0.05 was used to determine statistical significance.

3. Results

The fermentations were carried out at 28 °C for 7 days using different carbon sources and ingredients in the shake flask and fermentor.

The initial pH is one of the important parameters in the production of lycopene by *B. trispora*. The pH of the fermentation media changed during the fermentation time. The pH of the fermentation media was decreased during the first 2 days then increased during the remaining 7 days. At 2 days, the pH reduced to approximately 5.0, and then the pH increased over 6.0. The pH drop would be due to trisporic acid or/and other unidentified acidic formations in the lycopene production [17]. According to Nanou et al. [18], the pH of the media decreased slightly during the first 2 days of the fermentation from 7.6 to 5.3 and then increased slowly up to 7.0 at the end of the fermentation. The pH of the fermentation media was probably increased by ammonia liberation during degradation of proteins in the media by *B. trispora*. This would be indicative of amine metabolic product being released as a consequence of cell death and subsequent lysis occurring after that time.

3.1. Shake Flask Studies

Lycopene production of two glucose concentrations (4 and 6%) were investigated at two initial pH values (6.5 and 5.5). Figures 1 and 2 show the effect of glucose concentrations on lycopene production at the initial pH 6.5 and 5.5 respectively. Lycopene production at pH 6.5 with both glucose concentrations was higher than at pH 5.5. The amount of lycopene production was 77.7 and 28.1 mg L^{-1} with 4 and 6% glucose, respectively, after 7 days fermentation at pH 6.5. The biomass dry weight (DW) increased with the increase in glucose concentration from 4 to 6%. The lowest amount of lycopene content (0.3 mg g^{-1} DW) was obtained at pH 5.5 with glucose concentration 6%. In this research, 4% glucose yielded better lycopene than 6% while the biomass is better produced in 6% glucose. Since, there is an indirect relationship between biomass and lycopene concentration. Subsequently, increasing the glucose concentration to 6% resulted in a decrease in lycopene production by 36.2%. The decreased concentration of lycopene encountered with the highest concentration of glucose was probably due to osmotic effects. It has been reported that above a critical substrate concentration, the decreased water activity and the onset of plasmolysis combine to cause a decrease in the rates of fermentation and product formation [19]. In all cases, the biomass DW increased during the first two days of the fermentation and then remained constant until the maximum concentration of the lycopene was obtained. The maximum lycopene content was 2.4 and 0.6 mg g^{-1} DW with 4 and 6% glucose, respectively, at pH 6.5 after 7 days while they were 0.7 and 0.1 mg g^{-1} DW after 4 days, respectively. There are significant differences ($p < 0.05$) in lycopene production during fermentation. There are also significant differences ($p < 0.05$) in lycopene production between 4 and 6% glucose concentrations, and pH 6.5 and 5.5 at both glucose concentrations.

Figures 3 and 4 show the effect of sunflower and corn oils, respectively, on the production of lycopene in 4 % glucose medium at initial pH 6.5. The maximum lycopene was produced 795.1 and 714.2 mg L^{-1} with 1 % sunflower and corn oils respectively. The highest amount of lycopene (335.4 mg L^{-1}) was produced with 3 % sunflower oil rather 3 % corn oil (275.3 mg L^{-1}). Lycopene content was 131.4 and 115.0 mg g^{-1} DW at the end of fermentation for 1% sunflower and corn oils respectively. There are significant differences ($p < 0.05$) in lycopene production during the 7 days of fermentation. There are also significant differences ($p < 0.05$) in lycopene production with and without oil, and between both oil types.

Lycopene productions with 1 and 2% orange powder during 7 days of shake flask fermentation were given in Figure 5. Lycopene productions after 7 days were 2.9 and 1.7 mg L^{-1} with 1 and 2% orange powder respectively. Biomasses formations were 8.6 and 11.6 g L^{-1} for orange powder respectively. Lycopene content was 0.3 and 0.1 mg g^{-1} DW respectively at the end of fermentation. There were significant differences ($p < 0.05$) in lycopene production during the 7 days of fermentation for 1 and 2% orange powder, between the two orange powder concentrations.

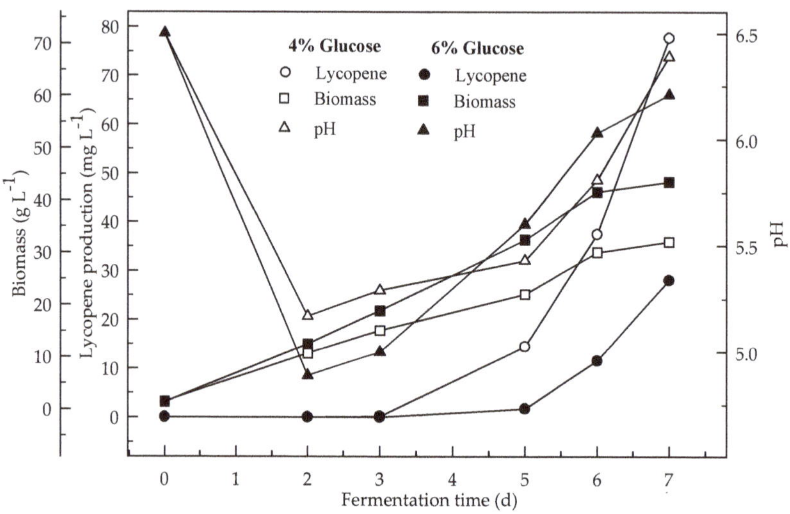

Figure 1. Changes in lycopene production, biomass formation and pH with initial pH 6.5 in a medium containing 4 and 6% glucose.

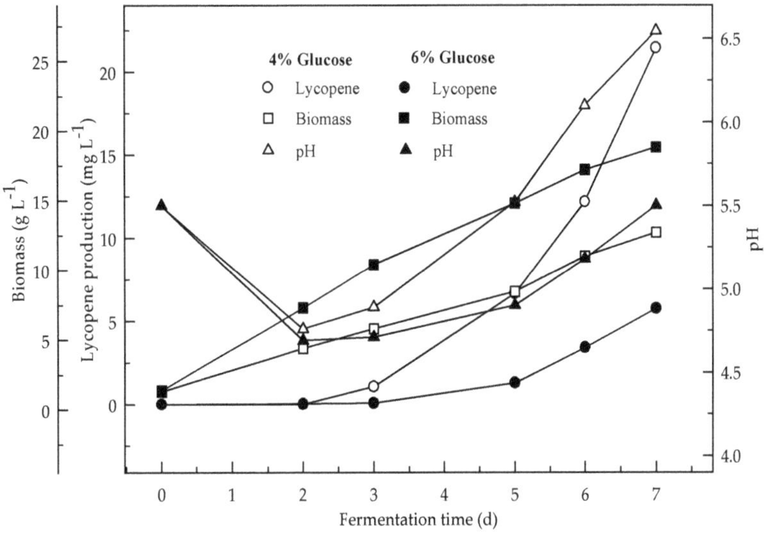

Figure 2. Changes in lycopene production, biomass formation and pH with initial pH 5.5 in a medium containing 4 and 6% glucose.

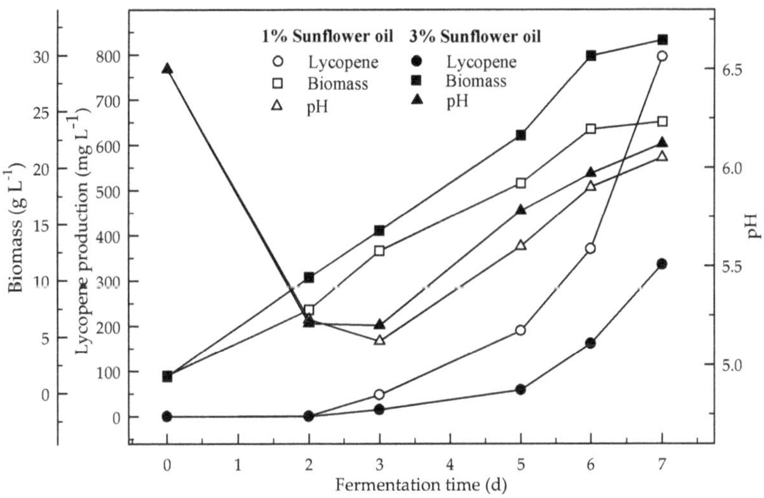

Figure 3. Changes in lycopene production, biomass formation and pH in fermentation with an initial pH 6.5 in a 4% glucose medium containing 1 and 3% sunflower oil.

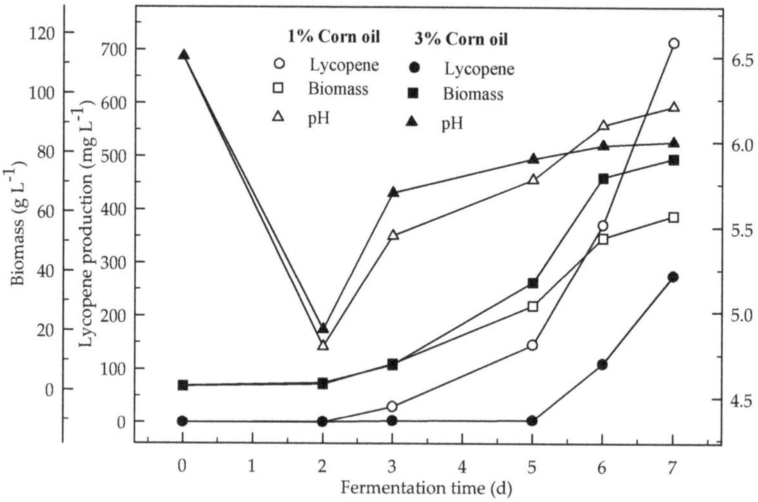

Figure 4. Changes in lycopene production, biomass formation and pH in fermentation with an initial pH 6.5 in a 4% glucose medium containing 1 and 3% corn oil.

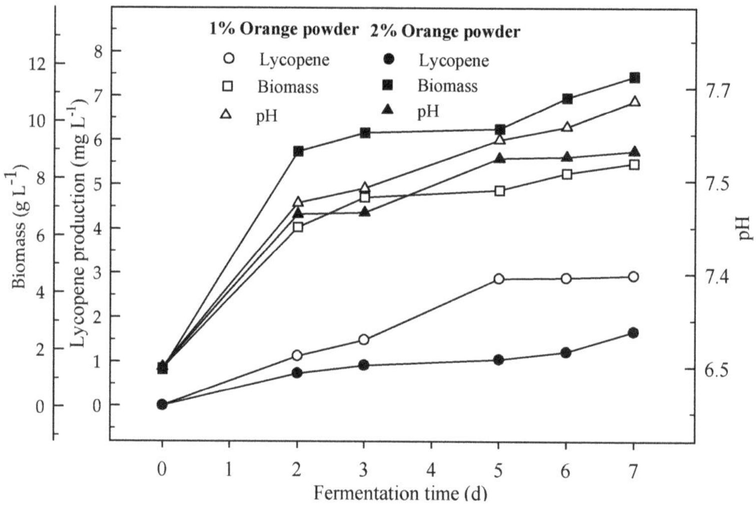

Figure 5. Changes in lycopene production, biomass formation and pH in a shake-flask fermentation with an initial pH 6.5 in a medium containing 1 and 2% orange powder.

3.2. Fermentor Studies

Effects of 4 and 6% glucose concentration (pH 6.5) on the production of lycopene were studied at 3 L min^{-1} aeration with 500 rpm agitation (Figure 6). The highest amount of lycopene was produced (92.3 mg L^{-1}) with 4% glucose concentration rather than 6% glucose (57.8 mg L^{-1}). Biomass DW formations after 7 days of fermentation were 154.9 and 196.7 mg L^{-1} with 4 and 6% glucose concentration respectively. Lycopene content was 3.8 and 1.9 mg g^{-1} DW at the end of fermentation for 4 and 6% glucose concentrations respectively. There is an indirect relationship between biomass and lycopene formation similar to the shake flask. The production of lycopene in the fermentor was higher than in the shake-flask from glucose. This would be due to the better aeration of the medium in the fermentor. There were significant differences ($p < 0.05$) in lycopene production during the 7 days of fermentation. There were also significant differences ($p < 0.05$) in lycopene production between 4 and 6% glucose concentrations.

The effect of high level of aeration (12 L min^{-1}) and agitation (750 rpm) on the production of lycopene in a 4% glucose medium during the 7 days of fermentation is given in Figure 7. The highest amount of lycopene was produced at low level of aeration (3 L min^{-1}) and agitation (500 rpm) rather than at the high levels (12 L min^{-1} and 750 rpm). The highest amount of biomass was formed at the higher aeration (12 L min^{-1}) and agitation (750 rpm) rather than at the lower levels. Lycopene content of DW (3.8 mg L^{-1}) was higher at 3 L min^{-1} aeration and 500 rpm agitation rather than 12 L min^{-1} at 750 rpm (1.8 mg L^{-1}). Generally, the results showed that the biomass increased with increasing aeration and agitation. This was due to the better air supply to the cells. This is especially important for high biomass concentrations. *B. trispora* is a strictly aerobic microorganism. Thus, moderate aeration rates in combination with high impeller agitation improved the growth of mold. There are significant differences ($p < 0.05$) in lycopene production during the 7 days of fermentation and between the two fermentation conditions (3 and 12 L min^{-1} aeration).

Figure 8 shows the effect of 1% sunflower and corn oils on the lycopene production in the fermentor. Lycopene production was 944.8 and 859.8 mg L^{-1} respectively. Biomass formations were 29.5 and 19.6 g L^{-1}. Lycopene content was 65.0 and 48.9 mg g^{-1} DW. Lycopene production in the presence of sunflower and corn oils in the fermentor was 12.2 and 11.1 times higher, respectively, then without oil from 4% glucose in a shake flask. This would be due to retaining higher amount of oxygen in the media. Since oil slightly increases the density of the fermentation media. There are significant differences ($p < 0.05$) in lycopene production during fermentation periods. There are also significant differences ($p < 0.05$) in lycopene production with and without oil, and between sunflower and corn oils.

Lycopene production was studied from different concentrations of orange peel in 4 L fermentation media with aeration 3 L min^{-1} and agitation 500 rpm (Figure 9). Lycopene was produced 4.9 mg L^{-1} with 1% orange peel. Biomass was 9.7 g L^{-1} and lycopene content was 0.05 mg g^{-1} DW at the end of fermentation. Lycopene production from orange peel was two times higher in the fermentor than in the shake flask. There are significant ($p < 0.05$) differences in lycopene production during the 7 days of fermentation.

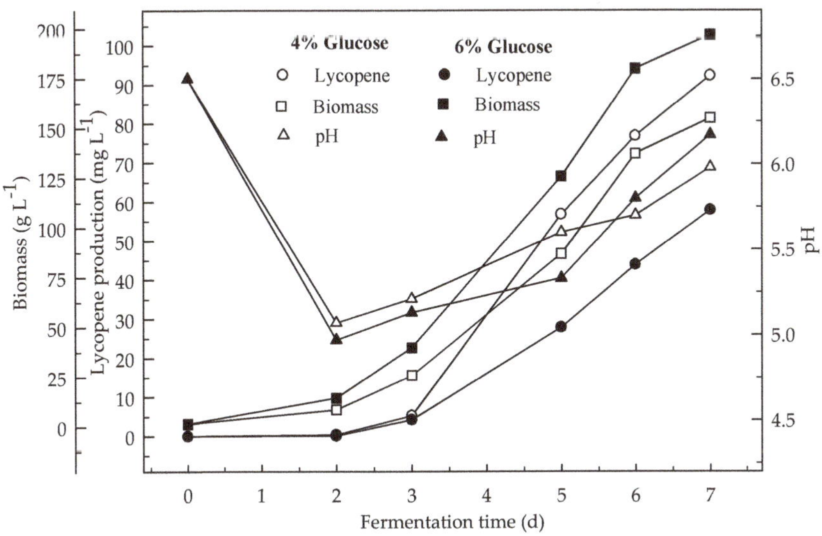

Figure 6. Changes in lycopene production, biomass formation and pH in a fermentor with an initial pH 6.5 in a 4 and 6% glucose medium at 3 L min^{-1} aeration and 500 rpm agitation.

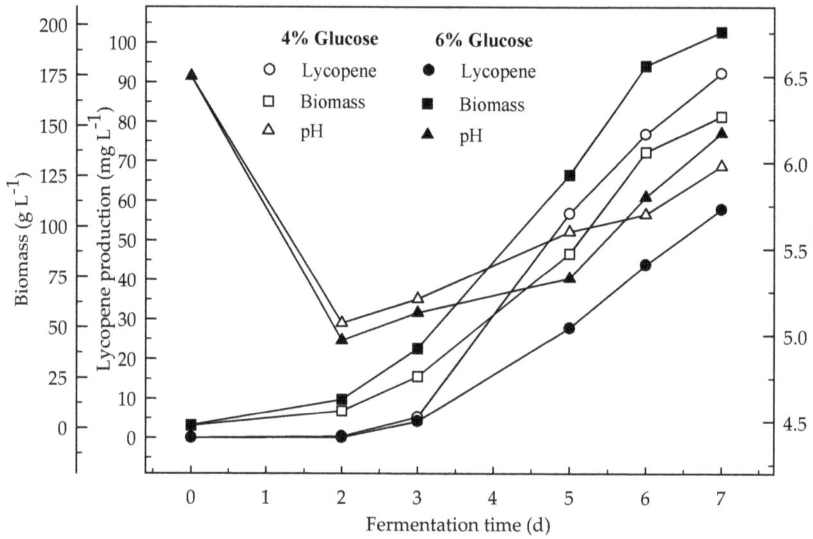

Figure 7. Changes in lycopene production, biomass formation and pH in a fermentor with an initial pH 6.5 in a 4 and 6% glucose medium at 12 L min^{-1} aeration and 750 rpm agitation.

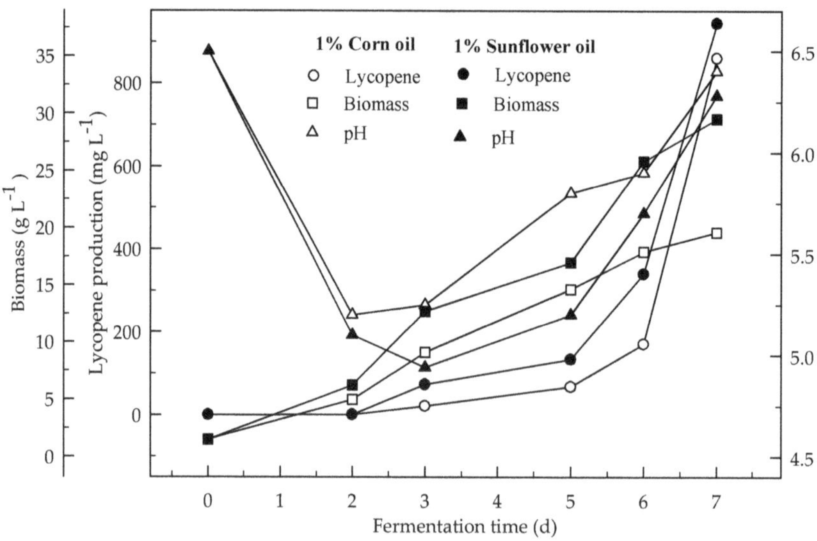

Figure 8. Changes in lycopene production, biomass formation and pH in a fermentor with an initial pH 6.5 in a 4% glucose fermentation medium 1% corn and sunflower oils (at aeration 3 L min^{-1} with 500 rpm agitation).

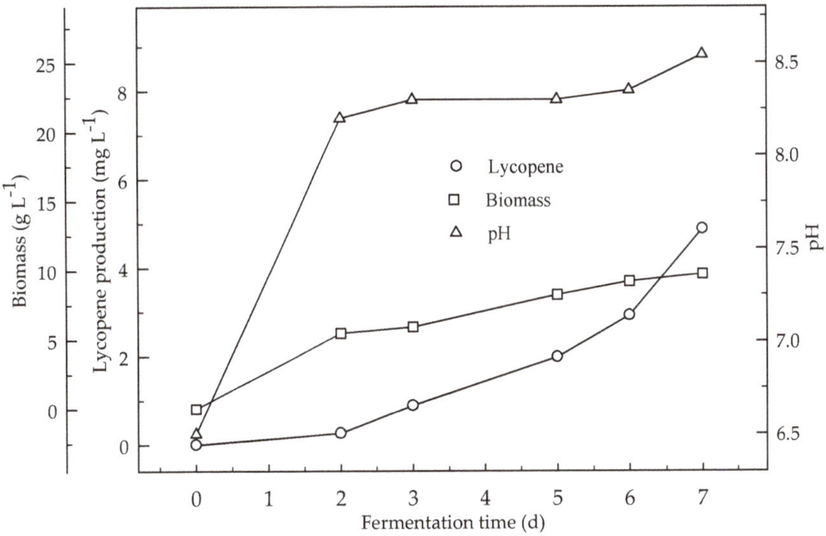

Figure 9. Changes in lycopene production, biomass formation and pH in a fermentor with an initial pH 6.5 from a 1% orange powder in a fermentor with aeration 3 L min^{-1} at agitation 500 rpm.

3.3. Morphological Analysis

The morphology of *B. trispora* was studied in the shake-flask and fermentor studies (3 L min^{-1} aeration and 500 rpm agitation) in only 4% glucose medium containing 1% sunflower oil using an image analysis system. During the first day of fermentation, microscopic examination showed that *B. trispora* formed a great amount of mycelium in the fermentation broth. After one day of fermentation, two opposite mating types of *B. trispora* produced zygospores which grew towards each other and produced pro-gametangia at their tips. Septation in the pro-gametangia led to the production of terminal gametangia, which fused to form the zygospores. The area of zygospores to the total area of mycelium remained almost constant up to 4 days of fermentation and then decreased significantly ($p < 0.05$) and disappeared on the 7th days of fermentation. On the other hand, the area of zygospores to the total area of mycelium increased significantly ($p < 0.05$) with the increase of fermentation time from 4 to 7 days. The hyphae length increased from the first day of fermentation up to 4 days and then slightly increased until the end of the fermentation. The diameter of zygospores increased from the first 2 days of incubation up to 4 days and then decreased slightly. Generally, morphological measurements using image analysis showed that *B. trispora* formed zygophores, zygospores and mycelium in which the hyphae appeared as a homogeneous dispersed suspension through the fermentation medium.

4. Discussion

Various concentration of carbon sources and refined natural vegetable oils as co-substrates were used in the production of lycopene. The concentration of carbon sources and vegetable oils greatly affected the final amount of lycopene. Furthermore, the use of orange powder components gives a first indication that *B. trispora* is able to metabolize and produce lycopene.

The highest concentration of lycopene was produced from a low concentration of glucose (4%) rather than at the high concentration (6%) in both the shake flask and fermentor studies. High concentrations of glucose may cause inhibition of the activity of mevalonate kinase [20], which is one of the key enzymes of the lycopene biosynthesis pathway. The carbon source can act as a major constituent for the building of cellular material and as an important energy source during the microbial fermentations [21]. Lycopene production in all fermentation conditions was lower than 1.0 mg L^{-1} during the first 2 days of fermentation and continuously increased during the next 5 days. It reached maximum concentration after 7 days. Sugar metabolism was slightly enhanced at the beginning of fermentation and increased during next fermentation period. This increase may provide more lycopene production [22]. On the other hand, biomass formation increased continuously during the 6 days of fermentation and it slowed on last day. When lycopene production increased, biomass formation decreased. Dry weight increased continuously up to 6 days and then slowed down. But lycopene production rate was higher on the last day. The maximum lycopene content in cells occurred after 7 days of fermentation when compared with 6 days in all fermentation conditions. This indicates that lycopene and biomass formations are inversely proportional. The biomass rapidly increased after 2 days of fermentation and this continued up to 6 days of fermentation. Nanou et al. [18] reported that the concentration of residual sugars fell rapidly during the first 4 days of the fermentation after which it decreased slowly and this was accompanied by a rapid increase of biomass concentration.

The refined vegetable oils stimulate different biosynthetic pathways in *B. trispora*. In this study, about 944.7 mg L^{-1} of lycopene was produced in fermentor studies from 4% glucose supplemented with 1 % sunflower oil. With 1% corn oil, about 859.8 mg L^{-1} of lycopene was produced in a 4% glucose medium. These results indicate that sunflower oil has a greater influence on lycopene production than corn oil due to a higher amount of vitamin E and linoleic acid in sunflower oil [23,24]. Similar results were demonstrated by Mantzouridou et al. [7] that corn steep liquor, olive oil, soybean oil, cottonseed oil and linoleic acids significantly increased the β-carotene production. Mantzouridou et al. [24] was also indicated that the addition of 10 g oil L^{-1} of substrate stimulated β-carotene production. Beside triacylglycerol, the main component of oils can be used in the lipid biosynthesis (involving acetyl-CoA formation resulting from β-oxidation of fatty acids) of the mold mycelium. Oils can be hydrolyzed to fatty acids and glycerol by mold exolipases [23]. When the concentration of natural oils increases (from 1 to 3%), the production of the lycopene decreases while the biomass dry weight increases significantly. This may be related to the consumed amount of the oils converted to biomass instead of lycopene. The production of lycopene without oils in the medium was lower while biomass formation was higher.

Two different aerations (3 and 12 L min^{-1}) and agitations (500 and 750 rpm) were studied in a 7 L fermentor (pH 6.5). The highest amount of lycopene 944.8 mg L^{-1} was produced with 3 L min^{-1} aeration and 500 rpm agitation in 4% glucose medium containing 1% sunflower oil. The positive impact of oils in the production of lycopene has been confirmed in other reports, such as those based on the use of crude olive or soybean olive [7,24] or waste cooked oil [25]. Supplementation of the medium with industrial glycerol, obtained either from soap manufacturing or biodiesel production industries, allows up to a tenfold increase in β-carotene levels [26]. Very promising results have been also obtained using agro-food wastes rich in carbohydrates and mineral salts, as beet molasses [27], cheese whey [28]. The concentration of lycopene was higher in fermentor studies than in the shake flask. Aeration could be beneficial to the growth and performance of aerobic *B. trispora*. Agitation of the fermentation medium creates shear forces which affect microorganisms in several ways: causing morphological changes, variation in their growth and product formation, and damaging the cell structure [19].

Orange powder was used as a raw material for production of lycopene in a shake flask and fermentor. The highest amount of lycopene (4.8 mg L^{-1}) was produced in the fermentor with 1% orange powder rather than in the shake flask (2.9 mg L^{-1}). Citrus peels are rich in cellulose, hemicelluloses, proteins and pectin, but the fat content is low. Pourbafrani et al. [21] reported that more than 95% of the peel oil is D-limonene which is extremely toxic to fermenting microorganisms. The addition of tomato sauce as a waste into the fermentation medium enhanced the production of lycopene and its production was 3.52 mg/100 g cell DW [29].

B. trispora is a micro-fungus with a life-cycle involving hyphae, zygophores and zygospores in submerged fermentation. The zygophores are the precursors of zygospores, which are responsible for the production of lycopene. [19]. The highest percentage of zygospores in an observed microscope field was correlated with the highest percentage of intracellular lycopene in the total biomass dry weight after the 7 days of fermentation. High percentages of vacuolated hyphae, evacuated cells and degenerated hyphae of *B. trispora* were observed at higher aeration (12 L min^{-1}). The percentage of the vacuolated hyphae, evacuated cells and the degenerated hyphae formed during the first 2 days of fermentation were higher than the remaining fermentation time. The growing mycelium was composed primarily of intact hyphae (such as 96.8% of the biomass dry weight), and zygospores on the 2nd day of fermentation, whereas on the next day the concentration of the intact hyphae was reduced to 67.9 of the biomass dry weight and the concentration of the zygospores was about 8.3%. On the 7th day of fermentation, the mycelium was composed mainly of evacuated cells. The addition of 1% oils to the fermentation medium increased the concentration of the intracellular lycopene during fermentation and reached to maximum level after 7 days, whereas the percentage of zygospores to the biomass dry weight increased significantly ($p < 0.05$). The highest concentration of lycopene correlated with the highest percentage of zygospores in the biomass DW. When the natural oils were used in the medium with 4% glucose in the fermentor, a great increase in the amount of lycopene occurred after the days 7 of fermentation, while the concentration of evacuated cells and the number of zygospores increased significantly ($p < 0.05$).

The effect of aeration rate and agitation speed on culture morphology was also observed. At an aeration rate of 3 L min^{-1} with 500 rpm agitation, the area of zygospores increased significantly and then remained constant. On the other hand, at aeration rates of 12 L min^{-1} with 750 rpm agitation, the area of zygospores decreased and then remained constant. The maximum concentration of lycopene was observed in the same conditions where a large number of zygospores were observed. This means that the zygospores were responsible for the synthesis of lycopene.

5. Conclusions

In this study, various substrates are compared with lycopene production in the shake flask and fermentor. The amount of lycopene production is based on the concentration of carbon sources and natural oil content. The amount of orange powder used in the fermentation as carbon sources affected the formation of lycopene. The use of natural oil is important for stimulation of the biosynthetic pathway in *B. trispora*. The medium with natural oil led to more lycopene than only carbon source in the medium. The oils degraded into acetyl-CoA, which is the precursor of lycopene derived from mevalonic acid. The mated fermentation process for lycopene production may be useful and a reference to the other fermentation. Also, lycopene production from *B. trispora* is better than other alternative plant extracts which is one of the other methods of lycopene production. For the future, this study is provided to contribute to the development in the lycopene production industry from this method. Oils improved lycopene production and enhanced vacuolation and evacuation of the hyphae. The morphological characteristics of the hyphae and the proportion of zygospores to the biomass DW were changed significantly ($p < 0.05$) within the medium composition and fermentation conditions. Moreover, there was a parallel relationship between the area of zygospores and production of lycopene. Generally, the results showed that *B. trispora* is a micro-mold that produces lycopene with a life-cycle involving hyphae, zygophores and zygospores. The hyphae do not play a role in the biosynthesis of lycopene. The zygphpores are the precursors of zygospores, which are responsible for the production of the lycopene. Maximum lycopene concentration was obtained when the medium was supplied with sunflower and corn oils.

Author Contributions: Examined the literature, methodology, original draft, writing-review and editing, supervision: O.E.; Performed the analysis, investigation, literature survey: A.S.

Funding: This research received no external funding.

Acknowledgments: The author acknowledges the support received from the Gaziantep University.

Conflicts of Interest: The author declares no conflict of interest.

References

1. Shi, Y.; Xin, X.; Yuan, Q. Improved lycopene production by *Blakeslea trispora* with isopentenyl compounds and metabolic precursors. *Biotechnol. Lett.* **2002**, *34*, 849–852. [CrossRef] [PubMed]
2. Liu, X.-J.; Liu, R.-S.; Li, H.M.; Tang, Y.-J. Lycopene production from synthetic medium by *Blakeslea trispora* NRRL 2895 (+) and 2896 (−) in a stirred-tank fermenter. *Biopro. Biosyst. Eng.* **2012**, *35*, 739–749.
3. Valduga, E.; Valério, A.; Treichel, H.; Júnior, A.F.; Di Luccio, M. Kinetic and stoichiometric parameters in the production of carotenoids by *Sporidiobolus salmonicolor* (CBS 2636) in synthetic and agroindustrial media. *Appl. Biochem. Biotechnol.* **2009**, *157*, 61–69. [PubMed]
4. Wang, J.F.; Liu, X.J.; Liu, R.S.; Li, H.M.; Tang, Y.J. Optimization of the mated fermentation process for the production of lycopene by *Blakeslea trispora* NRRL 2895 (+) and NRRL 2896 (−). *Biopro. Biosys. Eng.* **2012**, *35*, 553–564. [CrossRef] [PubMed]
5. Mehta, B.J.; Obraztsova, I.N.; Cerdá-Olmedo, E. Mutants and intersexual heterokaryons of *Blakeslea trispora* for production of beta-carotene and lycopene. *Appl. Env. Microbiol.* **2003**, *69*, 4043–4048. [CrossRef]
6. Tereshina, V.M.; Memorskaya, A.S.; Feovilova, E.P. Dormant cells in the development cycle of *Blakeslea trispora*: Distinct patterns of the lipid and carbohydrate composition. *Microbiology* **2002**, *71*, 684–689. [CrossRef]
7. Mantzouridou, F.; Roukas, T.; Kotzekidou, P.; Liakopoulou, M. Optimization of beta-carotene production from synthetic medium by *Blakeslea trispora*: A mathematical modeling. *Appl. Biochem. Biotechnol.* **2002**, *101*, 153–175. [CrossRef]
8. Tereshina, V.M.; Memorskaya, A.S.; Feovilova, E.P. Zygote formation in *Blakeslea trispora*: Morphological peculiarities and relationship with carotenoid synthesis. *Microbiology* **2003**, *72*, 448–454. [CrossRef]
9. Lopez-Nieto, M.J.; Costa, J.; Peiro, E.; Mendez, E.; Rodriguez-Saiz, M.; de la Fuente, J.L.; Cabri, W.; Barredo, J.L. Biotechnological lycopene production by mated fermentation of *Blakesiea trispora*. *Appl. Microbiol. Biotechnol.* **2004**, *66*, 153–159. [CrossRef]
10. Jeong, J.C.; Lee, J.; Park, Y.H. A unique pattern of mycelial elongation of *Blakeslea trispora* and its effect on morphological characteristics and beta-carotene synthesis. *Curr. Microbiol.* **2001**, *42*, 225–228. [CrossRef]
11. Stabnikova, O.; Wang, J.Y.; Ding, H.B.; Tay, J.H. Biotransformation of vegetable and fruit processing wastes into yeast biomass enriched with selenium. *Biores. Technol.* **2005**, *96*, 747–751. [CrossRef]
12. Bendini, A.; Lecce, G.D.; Valli, E.; Barbieri, S.; Tesini, F.; Toschi, T.G. Olive oil enriched in lycopene from tomato by-product through a co-milling process. *Int. J. Food Sci. Nutr.* **2015**, *66*, 371–377. [CrossRef] [PubMed]
13. Hernandez-Almanza, A.; Montanez, J.; Martínez, G.; Aguilar-Jiménezc, A.; Contreras-Esquivel, J.C.; Aguilar, C.N. Lycopene: Progress in microbial production. *Trend Food Sci. Technol.* **2016**, *56*, 142–148. [CrossRef]
14. Loco, J.V.; Js Elskens, M.; Croux, C.; Beernaert, H. Linearity of calibration curves: Use and misuse of the correlation coefficient. *Accred. Qual. Assur.* **2002**, *7*, 281–285.
15. Cucu, T.; Huvaere, K.; Bergh, M.-A.V.D.; Vinkx, C.; Loco, J.V. A simple and fast HPLC method to determine lycopene in foods. *Food Anal. Met.* **2012**, *5*, 1221–1228. [CrossRef]
16. Paul, G.C.; Thomas, C.R. Characterization of mycelial morphology using image analysis. *Adv. Biochem. Eng. Biotechnol.* **1998**, *60*, 1–59. [PubMed]
17. Papaioannou, E.H.; Liakopoulou-Kyriakides, M. Substrate contribution on carotenoids production in *Blakeslea trispora* cultivations. *Food Bioprod. Process.* **2010**, *88*, 305–311. [CrossRef]
18. Nanou, K.; Roukas, T.; Papadakis, E. Improved production of carotenes from synthetic medium by *Blakeslea trispora* in a bubble column reactor. *Biochem. Eng. J.* **2012**, *67*, 203–207. [CrossRef]
19. Mantzouridou, F.; Roukas, T.; Kotzekidou, P. Effect of the aeration rate and agitation speed on β-carotene production and morphology of *Blakeslea trispora* in a stirred tank reactor: Mathematical modeling. *Biochem. Eng. J.* **2002**, *10*, 123–135. [CrossRef]
20. Mantzouridou, F.; Roukas, T.; Kotzekidou, P. Production of beta-carotene from synthetic medium by *Blakeslea trispora* in fed-batch culture. *Food. Biotechnol.* **2004**, *18*, 343–361. [CrossRef]
21. Pourbafrani, M.; Talebnia, F.; Niklasson, C.; Taherzadeh, M.J. Effect of encapsulation in fermentation of limonene-containing media and orange-peel hydrolysate. *Inter. J. Molec. Sci.* **2007**, *8*, 777–787. [CrossRef]

22. Wang, H.-B.; Xu, R.-G.; Yu, L.-J.; Lu, J.; Zhang, L.-W.; Huang, X.-Y.; Zou, W.-A.; Zhao, Q.; Lu, M.-B. Improved β-carotene and lycopene production by *Blakeslea trispora* with ultrasonic treatment in submerged fermentation. *Z. Naturforsch. C* **2014**, *69*, 237–244. [CrossRef]
23. Vereschagin, O.A.; Memorskaya, A.S.; Tereshina, V.M. The role of exogenous lipids in lycopene synthesis in the Mucoraceous fungus *Blakeslea trispora*. *Microbiology* **2010**, *79*, 593–601. [CrossRef]
24. Mantzouridou, F.; Tsimidou, M.Z.; Roukas, T. Performance of crude olive pomace oil and soybean oil during carotenoid production by *Blakeslea trispora* in submerged fermentation. *J. Agric. Food Chem.* **2006**, *54*, 2575–2581. [CrossRef]
25. Nanou, K.; Roukas, T. Waste cooking oil: A new substrate for carotene production by *Blakeslea trispora* in submerged fermentation. *Biores. Technol.* **2016**, *203*, 198–203. [CrossRef]
26. Mantzouridou, F.; Naziri, E.; Tsimidou, M.Z. Industrial glycerol as a supplementary carbon source in the production of β-carotene by *Blakeslea trispora*. *J. Agric. Food Chem.* **2008**, *56*, 2668–2675. [CrossRef]
27. Goksungur, Y.; Mantzouridou, F.; Roukas, T.; Kotzekidou, P. Production of beta-carotene from beet molasses by *Blakeslea trispora* in stirred-tank and bubble column reactors: Development of a mathematical modeling. *Appl. Biochem. Biotechnol.* **2004**, *112*, 37–54. [CrossRef]
28. Papaioannou, E.H.; Liakopoulou-Kyriakides, M. Agro-food wastes utilization by *Blakeslea trispora* for carotenoids production. *Acta Biochim. Pol.* **2012**, *59*, 151–153. [CrossRef]
29. Radu, G.L.; Litescu, S.C.; Albu, C.; Truica, G. Beta-carotene and lycopene determination in new enriched bakery products by HPLC-DAD method. *Rom. Biotechnol. Lett.* **2012**, *17*, 7005–7012.

foods

Article

Assessment of Tryptophan, Tryptophan Ethylester, and Melatonin Derivatives in Red Wine by SPE-HPLC-FL and SPE-HPLC-MS Methods

Daniela Fracassetti [1,*], Ileana Vigentini [1], Alfredo Fabrizio Francesco Lo Faro [2], Patrizia De Nisi [3], Roberto Foschino [1], Antonio Tirelli [1], Marica Orioli [2] and Marcello Iriti [3]

1 Department of Food, Environmental and Nutritional Sciences, Università degli Studi di Milano, Via G. Celoria 2, 20133 Milan, Italy; ileana.vigentini@unimi.it (I.V.); roberto.foschino@unimi.it (R.F.); antonio.tirelli@unimi.it (A.T.)

2 Department of Biomedical, Surgical and Dental Sciences, Università degli Studi di Milano, Via Mangiagalli 37, 20133 Milan, Italy; fabriziolofaro09@gmail.com (A.F.F.L.F.); marica.orioli@unimi.it (M.O.)

3 Department of Agricultural and Environmental Sciences - Production, Landscape, Agroenergy, Università degli Studi di Milano, Via G. Celoria 2, 20133 Milan, Italy; patrizia.denisi@unimi.it (P.D.N.); marcello.iriti@unimi.it (M.I.)

* Correspondence: daniela.fracassetti@unimi.it; Tel.: +39-0250316674; Fax: +39-0250316672

Received: 22 February 2019; Accepted: 12 March 2019; Published: 14 March 2019

Abstract: Melatonin (MEL) is an indoleamine produced mainly by the pineal gland in vertebrates. It plays a significant role in the regulation of circadian rhythms, mitigation of sleeping disorders, and jet lag. This compound is synthetized from tryptophan (TRP) and it has been found in seeds, fruits, and fermented beverages, including wine. Wine is also a source of other tryptophan derivatives, the tryptophan ethylester (TEE) and MEL isomers (MISs), for which the biological properties need to be elucidated. An analytical method for the simultaneous quantification of TRP, TEE, and MEL was developed by a Solid Phase Extraction (SPE) of a preconcentration of wine followed by high performance liquid chromatography (HPLC) analysis either with fluorescence or mass spectrometer detectors. The analytical method showed a relative standard deviation (RSD) lower than 8%, except for TRP (RSD 10.5% in wine). The recovery was higher than 76%. The versatility of SPE preconcentrations allowed for the adequate preconcentration of wine sample as well as detection of low concentrations, an important aspect especially for MEL (detection limit 0.0023 µg/L). The proposed method proved to be suitable for assessing the investigated compounds in some red wine samples, where 74.4–256.2 µg/L and 0.038–0.063 µg/L of TEE and MEL were detected, respectively. Five MISs were also found in wine samples in concentrations up to 1.97 µg/L.

Keywords: indoleamines; grapevine; liquid chromatography; UHPLC/ESI-QTRAP; wine

1. Introduction

Melatonin (*N*-acetyl-5-methoxytryptamine; MEL) is an indoleamine produced mainly by the pineal gland in vertebrates [1] and its synthesis also occurs in other tissues, such as the gastrointestinal tract, bone marrow, and lymphocytes [2–4]. It is synthetized from L-tryptophan metabolism via serotonin [5,6]. In animals, MEL modulates the circadian and circannual rhythms, reproductive functions, bone metabolism, and turnover via cell-receptor-mediated mechanisms. It also shows a powerful antioxidant activity by directly scavenging free radical species (both reactive oxygen and nitrogen species) and stimulating the activity of antioxidant enzymes [7]. MEL has been found in edible plants [8–10] as a phytohormone protecting against oxidative stress and regulating growth [11,12]. Consequently, MEL is also present in plant foods, including seeds, fruits, and fermented beverages [11,13,14]. Several authors evidenced the presence of MEL in wines [15–18]. The circulating levels of MEL in mammals are very low (about 200 pg/mL at the maximum night peak and lower than 10 pg/mL during the day) [19] compared to MEL in grape products (about 1000 pg/g in berry skin and 500 pg/mL in wine) and, therefore, the intake of grape products represents a relevant topic [16,20]. The content of MEL increases during the fermentation step of the winemaking and other fermented beverages, meaning the yeast plays a significant role in its biosynthesis [14,21–23]. Besides MEL, its isomers (MISs) were also detected, and one of them was recently identified as tryptophan-ethyl ester (TEE) [24,25]. This compound was the most abundant tryptophan derivative, and its concentration was higher than that of MEL [26,27].

The analytical methods described for the detection of MEL in wine were based on liquid chromatography separation coupled with fluorescence [15,28,29] and mass spectrometry detectors [21,26,27]. Recently, Muñiz-Calvo and coauthors [30] proposed voltammetric techniques for the determination of MEL, TRP, and other TRP derivatives (TEE excluded). However, the authors evidenced the high performance liquid chromatography (HPLC) analysis had high discriminating capability and unambiguous identification and quantification. Due to the fact that wine is a complex matrix and MEL and its derivatives are present in low concentrations, the use of a solid phase extraction (SPE) technique is beneficial to use due to its removal of interferences, isolation, and concentration capabilities.

This study aimed to develop an HPLC method for the simultaneous determination of TRP, MEL, and TEE. In this respect, we developed a sample preparation protocol by means of SPE preconcentration allowing the detection of the three compounds by fluorescence and mass spectrometry detectors. The proposed method was used to determine the content of TRP, MEL, MISs, and TEE in red wine samples.

2. Materials and Methods

2.1. Chemicals and Materials

Melatonin (MEL), tryptophan-ethyl ester (TEE), methanol, and formic acid were purchased from Sigma-Aldrich (St. Louis, MO, USA). All the chemicals were of analytical grade. HPLC grade water was obtained by a Milli-Q system (Millipore Corp., Bedford, MA, USA).

MEL and TEE stock solutions were prepared in an ethanol:water ratio of 80:20 at 100 mg/L, while TRP was dissolved in water at 1000 mg/L. The stock solutions were stored covered with aluminum foil in order to protect the compounds against the light at −20 °C and they were thawed once.

Synthetic wine solution (SWS) contained 5.0 g/L tartaric acid and 12% ethanol (*v*/*v*) at a pH of 3.2, adjusted with sodium hydroxide (Merck, Darmstadt, Germany).

Eight red wine samples were analyzed. The wines were produced from *Vitis vinifera* cv. Nebbiolo in vintage 2015 in the Valtellina area (Lombardia, Italy), and they were collected at the market.

2.2. Sample Preparation

The preconcentration of red wine was developed by means of a solid phase extraction (SPE) technique. The SPE cartridges tested were Strata C18-T 500 mg/3 mL and Strata X-Polymeric Reversed Phase 200 mg/3 mL (Phenomenex, Torrance, CA, USA).

The final SPE protocol was set with Strata X-Polymeric Reversed Phase 200 mg/3 mL. After the SPE activation with methanol (4 mL) and conditioning with 0.1% formic acid (*v*/*v*) (8 mL), 5 mL of sample was loaded and the eluate was recovered (fraction A). The cartridge was firstly washed with 5 mL 0.1% formic acid in water (*v*/*v*) (fraction B) and then with 5 mL of 40% methanol (*v*/*v*) (fraction C). The elution was carried out with 5 mL of 100% methanol (fraction D), which was evaporated under vacuum (Rotavapor R 110, Büchi). The sample was re-suspended in 500 µL of 10% methanol acidified with 0.1% formic acid (*v*/*v*) corresponding to a concentration fold of 10. Each fraction was analyzed after filtration through a 0.22 µm polyvinilidene fluoride (PVDF) filter (Millipore). Samples were protected from the light during the preparation.

The setting of the SPE procedure was carried out on both spiked SWS and red wine at two concentration levels: 1 mg/L TRP, 50 μg/L TEE, and 50 μg/L MEL (level 1); and 2 mg/L TRP, 500 μg/L TEE, and 500 μg/L MEL (level 2). A higher concentration of TRP was chosen because its average content has been reported to be 3 mg/L [31]. The SPE procedure setup included differences in (i) volume of sample loaded (5 mL and 10 mL); (ii) volume of washing with 0.1% formic acid (*v*/*v*) (5 mL and 10 mL); (iii) volume and composition of washing with methanolic solution (5 mL and 10 mL; methanol 40%, 50%, 60%, and 70% (*v*/*v*)); (iv) recovery of fraction D with a different methanolic solution (methanol 50% (*v*/*v*), methanol 50% (*v*/*v*) acidified with 0.1% formic acid (*v*/*v*), 10% methanol (*v*/*v*), and 10% methanol (*v*/*v*)acidified with 0.1% formic acid (*v*/*v*)).

2.3. Chromatographic and Quantification Conditions

The HPLC system consisted of an Agilent 1260 Infinity Quaternary LC equipped with an Agilent G1321B Fluorescence Detector and an Agilent LC/MS 6130 Quadrupole (Agilent, Santa Clara, CA, USA) operating in multimode source (simultaneous electrospray (ESI) and atmospheric pressure chemical ionization (APCI)). The MS operative conditions included a spray voltage of +3.5 kV, vaporizer temperature of 250 °C, gas temperature of 350 °C, drying gas at 12 mL/min, and a nebulizer pressure of 55 psig. The fluorescence detector was set at 280 nm and 350 nm for excitation and emission, respectively. The column used for the separation was an Accucore C18 (100 × 3 mm, 2.7 μm particle size, Thermo Scientific, San Jose, CA, USA) set at 40 °C. The elution solvents were: (A) 0.1% formic acid (*v*/*v*), and (B) methanol acidified with 0.1% formic acid (*v*/*v*), and the flow rate was 0.5 mL/min. Elution conditions were as follows: 5% B for 1 min, from 5% to 40% B in 15 min followed by the column washing with 100% B and re-equilibration for a total of 25 min for each run. Injection volumes were 5 μL and 20 μL for HPLC-MS and HPLC-FL methods, respectively.

Chromatographic data were processed using Agilent OpenLab ChemStation software (Agilent). For the MS detector, data acquisition was performed in a single ion monitoring (SIM) setting [m/z]+ 205 for TRP and 233 [m/z]+ for MEL and TEE. MEL, TEE, and TRP were quantified using the external standard method. Six-level calibration curves were obtained with standard solutions containing the analytes at the respective concentrations spanning the expected ranges in wine.

The UHPLC/ESI-QTRAP analysis was carried out to identify MEL, MISs, and TEE eluting in the chromatograms of the wine samples. A UHPLC model HP 1290 (Agilent) coupled with a Mass Spectrometer Detector 5500 TRIPLE QTrap model 1024945-AX (ABSCIEX, Framingham, MA, USA) equipped with an HESI-II probe for electrospray ionization and a collision cell (HCD) was used. Data were acquired in a multiple reaction monitoring (MRM) mode. The operative conditions were: spray voltage +2.2 kV, sheath gas flow-rate 50, auxiliary gas flow-rate 55, capillary temperature 360 °C, capillary +95 V, tube lens +170 V, Skimmer +38 V, and heater temperature 500 °C. The chromatographic separation was carried out on a BEH C18 column (100 × 2.1 mm, 1.7 μm particle size, Waters, Milford, MA, USA). The column was maintained at 30 °C and the injection volume was 0.3 μL. The elution solvents were (A) 0.1% formic acid (*v*/*v*), and (B) acetonitrile. The elution conditions were as follows: 10% B for 0.30 min, from 10% to 30% B in 5 min, from 30% to 85% in 0.5 min followed by column washing with 90% B and re-equilibration for a total of 10.5 min for each run. The flow rate was set at 0.55 mL/min. The MS data were processed using MultiQuant software (AB Sciex). The peak identity was ascertained by evaluation of both the accurate mass and the fragments obtained in the collision cell. MEL, TEE, and TRP were quantified using the external standard method with the respective standards; MISs were quantified with the calibration curve obtained for MEL.

2.4. Method Validation

The in-house validation of the method was carried out in terms of selectivity, linearity, limit of detection (LOD), limit of quantification (LOQ), repeatability, and recovery [32].

The selectivity of the method was evaluated by analyzing TRP, TEE, and MEL both in the absence and presence of possible interferences originating from the wine matrix. With this aim, the separations of a SWS at the concentrations 2 mg/L TRP, 50 μg/L TEE, and 50 μg/L MEL analyzed with both FL and MS detectors were compared with those of a purified SWS and a red wine sample spiked with the three analytes at the same concentrations.

Linearity was tested on six concentration levels within the intervals reported in Tables 1 and 2. The SWS samples were purified and analyzed in triplicate as well as spiked red wine samples at the same concentration levels. The equations of the calibration curves and the correlation coefficients (r) were obtained by linear regression analysis.

The values of limit of detection (LOD) and limit of quantification (LOQ) were calculated as the lowest concentration of analyte in a sample that resulted in a signal-to-noise ratio of 3 and 10 for LOD and LOQ, respectively. Values were measured in three independent replicates.

The recovery (%) in purified SWSs and spiked wine samples was calculated using the SWSs added with the same concentrations of the analytes. The spiking concentrations were 5.3 μg/L, 10.6 μg/L, 21.2 μg/L, 53.0 μg/L, 106.0 μg/L, and 212 μg/L for TEE and MEL, and 110 μg/L, 220 μg/L, 550 μg/L, 1100 μg/L, 2200 μg/L, and 5500 μg/L for TRP. SWSs and spiked red wine samples were purified and analyzed in triplicate. Triplicate analyses of unpurified SWSs spiked with the same concentrations of the analytes were also carried out. The accuracy was evaluated by means of recovery assay of replicate preconcentrations of both SWS and spiked red wine samples at levels of 10.6 μg/L and 106.0 μg/L for TEE and MEL, and 220 μg/L and 2200 μg/L for TRP. Duplicate preconcentrations were carried out in two different days for SWSs (n = 4) and for three different days for spiked red wine samples (n = 6).

Precision was expressed as relative standard deviation (RSD) of the analytical response. Both the SWSs and red wine samples spiked at the three levels of concentration were analyzed (10.6 μg/L, 53.0 μg/L, and 106.0 μg/L for TEE and MEL, and 220 μg/L 1100 μg/L, and 2200 μg/L for TRP). Three independent replicates were carried out on three different days (n = 9). For the repeatability of the SPE procedure and instrumental repeatability, duplicate injections were carried out from the same vial at the two concentration levels on three different days (n = 12). The repeatability of the retention times was also evaluated over a total of 54 injections carried out on three different days.

2.5. Statistical Analysis

Statistical analysis was carried out with SPSS Win 12.0 program (SPSS Inc., Chicago, IL, USA). The equations of the calibration curves were assessed by the linear regression analysis. Differences between the calibration curve slopes obtained in aqueous solution and white wine were evaluated by the F-test ($p < 0.05$).

3. Results and Discussion

3.1. Analytical Method Development

The chromatographic conditions by means of both FL and MS detectors were firstly optimized using the SWS at the concentration 2 mg/L TRP, 50 μg/L TEE, and 50 μg/L MEL. While the analytical response (area) was similar for MEL and TEE using MS detector, it resulted lower in fluorescence for TEE analyzed at the same concentration (Figure 1).

A standard water solution with the same levels of analytes was analyzed, and no differences were observed in comparison to SWS. The SPE preconcentration allowed both the removal of compounds potentially interfering in the HPLC separation from the wine matrix and the concentration of the analytes of interest, making their detection possible because of their presence in low amounts in wine [33]. Although other authors applied the SPE preconcentration for MEL detection from wine and other food matrices [15,27,29,34], the SPE approach presented in this study allowed the simultaneous elution of TRP, TEE, and MEL. For the development of the SPE preconcentration, several assays were carried out using the SWS containing concentrations of 2 mg/L TRP, 50 μg/L TEE, and 50 μg/L MEL. The setup method by using a blank matrix added with the compounds of interest is considered acceptable [35] since a Certified Reference Material for wine containing TRP, TEE, and MEL does not exist. First, preconcentrations by using two different SPE cartridges were compared by loading 10 mL of spiked SWS and washing with methanol 40% methanol (*v*/*v*). Higher recovery (>15%) was observed by the preconcentration with Strata X-Polymeric Reversed Phase 200 mg, in accordance to El-Moussaoui and Bendriss [29]. These SPE cartridges were used for the development of the analytical method. Trials loading 5 mL and 10 mL of spiked SWS at two concentration levels (see Paragraph 2.2), and washing with 5 mL and 10 mL 0.1% formic acid (*v*/*v*), and 5 mL 40% methanol (*v*/*v*), led to comparable recovery values. The lower amounts were chosen in order to reduce the solvent amounts needed for the SPE, which were 5 mL of 0.1% formic acid and 5 mL of methanol in total for the four steps of preconcentration. The analyses of all these fractions were carried out. TRP was detected in the sample loading (fraction A) and in the washing with 40% methanol (fraction C), TEE was revealed in fraction C and in the methanolic solution (fraction D), while MEL was only found in fraction D. The major interferences of red wine are represented by phenols, compounds completely soluble in methanol. In order to reduce the possible interferences in fraction D, different concentrations of methanol (40–70% (*v*/*v*)) were assayed. However, the increase of methanol in the washing step led to a loss of MEL in fraction D, even when the washing was carried out with 50% methanol (*v*/*v*). The recovery of fraction D, after methanol evaporation under vacuum, was carried out with different methanol amounts (500 µL and 1 mL) and concentrations (10% and 50% (*v*/*v*)), with and without acidification with 0.1% formic acid (*v*/*v*). The best chromatographic separation in terms of peak shape was obtained with methanol 10% acidified with 0.1% formic acid (*v*/*v*). Formic acid 0.1% (*v*/*v*) was the solvent for the HPLC separation and no change of sample pH was generated through the analysis.

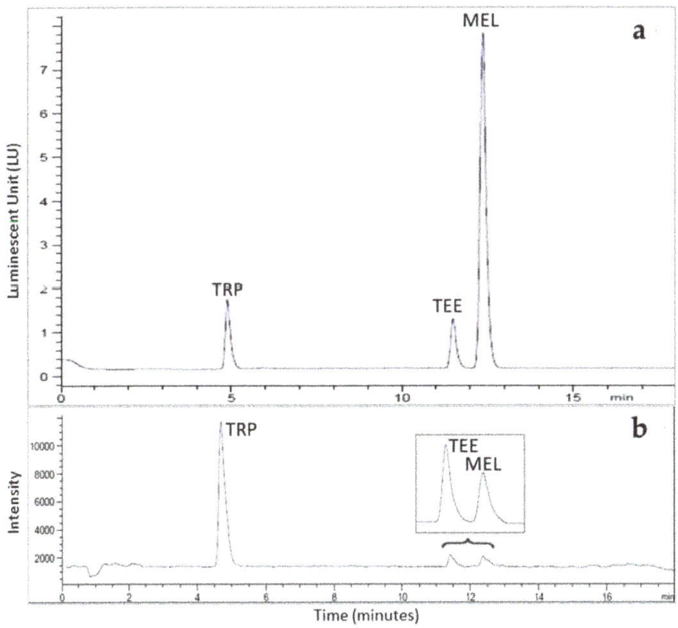

Figure 1. (**a**) HPLC-FL and (**b**) HPLC-MS chromatograms of tryptophan (TRP) (1 mg/L), tryptophan ethyl ester (TEE) (50 µg/L), and melatonin (MEL) (50 µg/L) in standard synthetic wine solution.

Based on the results obtained for the spiked SWS, the SPE procedure was carried out by loading 5 mL of spiked SWS and spiked wine sample recovering the eluate (fraction A). The cartridge was firstly washed with 5 mL of 0.1% formic acid (*v*/*v*) (fraction B) and then with 5 mL of 40% methanol (*v*/*v*) (fraction C). The elution was carried out with methanol 100% (fraction D) which was evaporated under vacuum. The fraction D was re-suspended in 500 µL of 10% methanol acidified with 0.1% formic acid (v/v), meaning a 10-fold concentration of wine sample.

Before proceeding to method validation, spiked red wine samples at two concentration levels (see Paragraph 2.2) were SPE preconcentrated in order exclude any possible interferences arising from the HPLC separation of red wine, especially for the concentrated fraction D. As shown in Figure 2, the HPLC separation was interference-free, not only for fraction D, but also for fractions A and C.

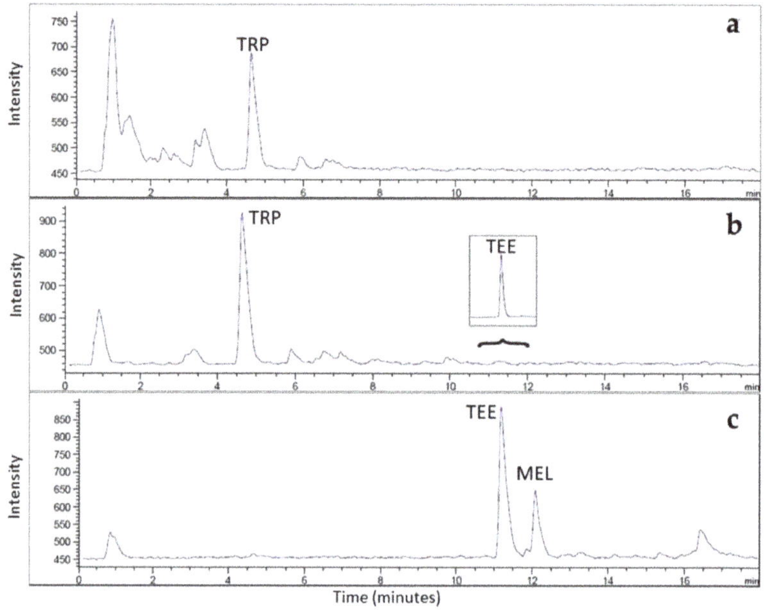

Figure 2. HPLC-MS chromatograms of (**a**) fraction A, (**b**) fraction C, and (**c**) fraction D obtained for solid phase extraction (SPE)-preconcentrated red wine sample spiked with TRP (1 mg/L), TEE (50 µg/L), and MEL (50 µg/L). Peak identity: TRP, tryptophan; TEE, tryptophan ethyl ester; MEL, melatonin.

3.2. Validation of the Analytical Method

The linearity was firstly assessed for TRP, TEE, and MEL added to SWS by calculating six-point calibration curves without SPE preconcentrations. Both the FL and MS detectors showed correlation coefficient (r) values higher than 0.99 (Table 1). A linear response was also found for TRP, TEE, and MEL determined in spiked SWSs after SPE preconcentration and analyzed by the FL detector ($y = 39.03 \times x + 0.07$ for TRP, $y = 0.51 \times x + 58.46$ for TEE, and $y = 1.19 \times x - 9.00$ for MEL) with r values that corresponded to 1, 0.991, and 0.998 for TRP, TEE, and MEL, respectively. The LOD and LOQ were assessed for the three compounds and they resulted in lower than the average concentrations found in wine [21,28,33,36] for both TRP (5.16 µg/L and 17.18 µg/L for LOD and LOQ, respectively) and 22 MEL (0.12 µg/L and 0.41 µg/L for LOD and LOQ, respectively). Higher LOD and LOQ were found for TEE (1.17 µg/L and 3.91 µg/L, respectively) making the analytical method proposed less sensitive for TEE quantification in comparison to MEL. However, Vigentini et al. [26] and Fernández-Cruz et al. [27] reported concentrations of TEE higher than the LOQ value; the mentioned authors found TEE amounts in most cases higher than 5 µg/L and up to 1126 µg/L.

A linear response was observed for both spiked SWSs and red wine samples analyzed by the MS detector, and no significant difference was found between the two matrices. The r values were higher than 0.99 for all the compounds investigated, and no significant differences were found in the calibration curve slopes calculated in spiked SWSs and red wine samples purified with SPE (Table 2).

As expected, both LOD and LOQ values were lower when the detection was carried out by the MS detector (Table 2), which is more sensitive than FL detector. The LOD values we found were on the same order of magnitude as other analytical methods previously proposed for TEE, while the results were lower in the case of MEL, indicating a high sensitivity of the analytical method proposed [26–28,37]. In addition, the chromatographic separation was interference-free (Figure 2), facilitating the identification and quantification.

Table 1. Linearity obtained with standard synthetic wine solutions by fluorescence (FLD) and mass spectrometry (MSD) detectors.

	FLD			MSD		
Compound	Concentration Range (μg/L)	Linearity		Concentration Range (μg/L)	Linearity	
		Equation	*r*		Equation	*r*
TRP	110–5500	$39.03 \times x - 0.08$	0.996	110–5500	$3389.30 \times x + 49.99$	1
TEE	50–2000	$0.064 \times x + 7.308$	0.998	5–200	$23.40 \times x + 111.85$	0.999
MEL	20–500	$0.15 \times x + 0.41$	1	1–200	$15.04 \times x + 109.10$	0.998

The equations and the correlation coefficients (*r*) were calculated by means of linear regression. *x*, concentration; TRP, tryptophan; TEE, tryptophan ethyl ester; MEL, melatonin.

The recovery of the SPE preconcentration was evaluated for both spiked SWSs and red wine samples at six concentration levels (see Paragraph 2.4). The average values of percent recovery were all higher than 75% (Table 2). Only negligible differences were obtained by analyzing the spiked SWSs and red wine samples, thus a matrix effect could be excluded (Table 2).

The repeatability of the method was calculated as the relative standard deviation (RSD) of the analytical response (peak area) obtained by analyzing both spiked SWSs and red wine samples ($n = 9$). The average RSD values obtained for the TRP, TEE, and MEL in spiked SWSs ranged between 4.6% and 9.1%, whereas values obtained for spiked red wine samples were between 5.4% and 10.5% (Table 2). The repeatability of the SPE preconcentration was also evaluated for two concentration levels in both spiked SWS ($n = 4$) and red wine ($n = 6$). The average values determined for spiked SWS were 5.4%, 7.4%, and 4.0%for TRP, TEE, and MEL, respectively, and the values for spiked red wine samples were 4.2%, 5.2%, and 5.2% for TRP, TEE, and MEL, respectively. The intra-day precision ($n = 12$) corresponded to 3.8%, 2.6%, and 3.1% for TRP, TEE, and MEL, respectively. The repeatability of retention times (n = 54), estimated on three different days, was negligible (<2%). Overall, these data confirm the reliability of the analytical method here proposed.

The last step of the developed sample preparation method was the evaporation of methanol under vacuum and the re-suspension of samples in 500 μL of 10% methanol acidified with 0.1% formic acid (*v*/*v*), meaning a 10-fold concentration of wine sample. Assays were carried out by concentrating the wine samples up to 25-fold, and results were in accordance to those found for 10-fold concentrated samples (data not shown). These findings indicated the high versatility of the developed SPE preconcentration, making possible the detection of the investigated compounds, MEL and TEE in particular, even when they are present in very low concentrations (<0.02 μg/L).

3.3. Analysis of Red Wine Samples

In order to evaluate the suitability of the analytical method proposed, the determinations of TRP, TEE, and MEL were carried out in eight red wine samples produced from *Vitis vinifera* cv. Nebbiolo in vintage 2015. The wine samples were also analyzed by UHPLC/ESI-QTRAP in order to confirm the presence of MEL and TEE. Besides the latter two compounds, five MISs were detected, and fragmentation is reported in Table 3. This is an improvement of the existing methods since one MIs [38,39] to three MISs [26,33] were previously detected in wine.

Table 2. Linearity, limits of detection (LOD) and quantification (LOQ), recovery (%), and repeatability (as relative standard deviation, %RSD) for the analytical method developed in HPLC-MS.

Compound	Concentration Range Added (μg/L)	SWS		Spiked Red Wine		LOD (μg/L) (n = 3)	LOQ (μg/L) (n = 3)	Recovery (%) (n = 6)		Repeatability (%RSD) (n = 9)	
		Equation	r	Equation	r			SWS	Spiked Red Wine	SWS	Spiked Red Wine
TRP	110–5500	$3284.2 \times x + 328.3$	0.999	$3257.4 \times x + 3132.1$	0.995	0.75	1.25	89	84	9.1	10.5
TEE	5–250	$659.2 \times x - 440.4$	0.997	$677.4 \times x + 3861.0$	0.999	0.038	0.12	88	76	6.5	7.9
MEL	0.05–250	$242.2 \times x + 2748.6$	0.996	$269.5 \times x + 840.1$	0.997	0.0023	0.018	86	79	4.6	5.4

The equations and the correlation coefficients (r) were calculated by means of linear regression. Recovery and repeatability values were determined at two and three different concentration levels, respectively. SWS, synthetic wine solution; x, concentration; TRP, tryptophan; TEE, tryptophan ethyl ester; MEL, melatonin.

Table 3. Identification of MEL (melatonin), TEE (tryptophan ethyl ester), and MISs (MEL isomers) by HUPLC/ESI-QTRAP.

Compound	Exact Mass	MS/MS Fragmentation	
	$[M + H]^+$	MS/MS Fragments	Collision Energy (μV)
MEL	233.1	188.1	30
		216.1	30
		174.1	30
TEE	233.1	174.1	20
		159.0	36
		130.1	55
		178.1	29
MIS 1	233.1	141.0	20
		216.0	20
		174.1	20
MIS 2	233.1	141.1	20
		196.0	20
MIS 3	233.1	130.0	50
MIS 4	233.1	141.0	20
		216.0	20
		174.1	20
MIS 5	233.1	141.0	35
		159.0	35

The concentration of TRP varied from 0.44 ± 0.05 mg/L to 4.39 ± 0.46 mg/L (Table 4); the concentration of free amino acids can be quite variable in wine, depending on the initial content in grapes and cellular lysis occurrence of wine-related microorganisms [31]. The revealed TEE was variable among wine samples, ranging from 74.4 ± 5.9 μg/L to 256.2 ± 20.2 μg/L (Table 4). These amounts were higher in comparison to the values reported by Vigentini et al. [26] and Fernández-Cruz et al. [27] who monitored the TEE content in laboratory-scale fermentations in real must and in synthetic must, respectively. Maybe winemaking at an industrial-scale or the unknown fermenting yeast could affect the level of TEE. MEL was detected in the analyzed red wine samples, ranging from 0.038 ± 0.001 μg/L to 0.063 ± 0.004 μg/L (Table 4), on the same order of magnitude as the concentration reported by Vitalini and co-authors [33] who detected MEL in red wines at 0.05–0.62 μg/L. These amounts could be of physiological interest since the MEL detected in wine was higher than its circulating levels in humans during the day (10 pg/mL) [19].

Table 4. Levels of TRP (tryptophan), TEE (tryptophan ethyl ester), MEL (melatonin), and MEL isomers (MISs) in red wine samples. Data are expressed as mean ± standard deviation.

Sample Code	TRP	TEE	MEL	MIS 1	MIS 2	MIS 3	MIS 4	MIS 5
	mg/L	μg/L	μg/L	μg/L	μg/L	μg/L	μg/L	μg/L
Red wine 1	3.85 ± 0.40	172.2 ± 13.6	0.057 ± 0.003	1.64 ± 0.09	<LOQ	<LOQ	0.0043 ± 0.0002	<LOQ
Red wine 2	4.39 ± 0.46	212.0 ± 16.8	0.062 ± 0.003	1.97 ± 0.11	<LOQ	<LOQ	0.0041 ± 0.0002	<LOQ
Red wine 3	1.56 ± 0.16	256.2 ± 20.2	0.063 ± 0.004	0.74 ± 0.04	<LOQ	<LOQ	<LOQ	<LOQ
Red wine 4	1.02 ± 0.11	223.2 ± 17.6	0.038 ± 0.002	0.67 ± 0.04	<LOQ	<LOQ	<LOQ	<LOQ
Red wine 5	0.98 ± 0.10	113.0 ± 8.9	0.046 ± 0.003	0.58 ± 0.03	<LOQ	<LOQ	<LOQ	<LOQ
Red wine 6	0.84 ± 0.09	92.9 ± 7.3	0.054 ± 0.003	0.86 ± 0.05	<LOQ	<LOQ	<LOQ	<LOQ
Red wine 7	0.44 ± 0.05	71.7 ± 5.7	0.063 ± 0.004	0.91 ± 0.04	<LOQ	<LOQ	<LOQ	<LOQ
Red wine 8	0.57 ± 0.06	74.4 ± 5.9	0.038 ± 0.001	0.80 ± 0.04	<LOQ	<LOQ	<LOQ	<LOQ

Several factors can affect the concentration of MEL in red wine, including agrochemicals (i.e., plant activators, treatment with copper) and winemaking practices (i.e., barrel-aging, content of sulphur dioxide, and the use of fining agents), fermenting microorganisms (different strains of both *Saccharomyces cerevisiae* and non-*Saccharomyces* yeasts), and maybe even the grape/wine composition (i.e., phenol content, antioxidant capacity) [18,26]. However, further investigations will be carried out in order to elucidate the effects of both winemaking and grape/wine composition. Most of the MISs were detected at trace levels except for MIS 1, for which the concentrations ranged from 0.58 ± 0.03 μg/L to 1.97 ± 0.11 μg/L (Table 4), more than one order of magnitude higher in comparison to MEL. MIS biosynthesis is still not clear, and their antioxidant and cytoprotective activities are dependent on the position of two side chains in the indole ring [40].

4. Conclusions

The proposed analytical method allowed the detection and reliable quantification of TRP, TEE, and MEL simultaneously. In general, this protocol may represent a useful tool for monitoring the release of MEL and TEE, even when their concentration is very low (<0.02 μg/L), and their fate throughout the wine production and storage. Moreover, the sample preparation method and chromatographic conditions presented in this study allowed the detection of five MISs, for which the molecular structure, in particular the position of two side chains in the indole ring, will be evaluated.

The obtained results suggested the amounts of MEL detected in the wine samples may be of biological importance, as MEL was higher than its circulating levels in humans during the day; bioavailability studies are necessary to confirm this hypothesis. TEE is more highly concentrated than MEL (over one thousand times) in Nebbiolo wine samples, as well as MIS 1, confirming the outcome of previous works [24,26], even though the origin and the putative nutritional role of this molecule have not been elucidated yet. Since chromatographic analysis does not allow getting information about the chemical structure of MISs, further investigations will be carried out in order to clarify the position of two side chains in the indole ring for the MISs detected. We can suppose that the proper choice of the winemaking procedures could lead to higher levels of MEL, TEE, and MISs, and this needs further investigation.

Author Contributions: Conceptualization, D.F., M.I. and I.V.; Methodology, D.F.; Validation, D.F., A.F.F.L.F., P.D.N.; Data Curation, D.F.; Resources, R.F., M.I., M.O. and A.T.; Writing—Original Draft Preparation, D.F. and I.V.; Writing—Review & Editing, P.D.N., Roberto Foschino, M.I., A.F.F.L.F., M.O. and A.T.

Funding: This research received no external funding.

Acknowledgments: We are grateful to Sonia Moiola and Alice Boletti for their technical support.

Conflicts of Interest: The authors declare no conflict of interest.

References

1. Reiter, R.J. The pineal and its hormones in the control of reproduction in mammals. *Endocr. Rev.* **1980**, *1*, 109–131. [CrossRef] [PubMed]
2. Huether, G.; Poeggeler, B.; Reimer, A.; George, A. Effect of tryptophan administration on circulating melatonin levels in chicks and rats: Evidence for stimulation of melatonin synthesis and release in the gastrointestinal tract. *Life Sci.* **1992**, *51*, 945–953. [CrossRef]
3. Huether, G. The contribution of extrapineal sites of melatonin synthesis to circulating melatonin levels in higher vertebrates. *Experientia* **1993**, *49*, 665–670. [CrossRef] [PubMed]
4. Tan, D.X.; Manchester, L.C.; Terron, M.P.; Flores, L.J.; Reiter, R.J. One molecule, many derivatives: A never-ending interaction of melatonin with reactive oxygen and nitrogen species? *J. Pineal Res.* **2007**, *42*, 28–42. [CrossRef] [PubMed]
5. Reiter, R.J. Pineal melatonin—Cell biology of its synthesis and of its physiological interactions. *Endocr. Rev.* **1991**, *12*, 151–180. [CrossRef]
6. Chattoraj, A.; Liu, T.; Zhang, L.S.; Huang, Z.; Borjigin, J. Melatonin formation in mammals: In vivo perspectives. *Rev. Endocr. Metab. Disord.* **2009**, *10*, 37–243. [CrossRef]
7. Reiter, R.J.; Paredes, S.D.; Manchester, L.C.; Tan, D.X. Reducing oxidative/nitrosative stress: A newly-discovered genre for melatonin. *Crit. Rev. Biochem. Mol. Biol.* **2009**, *44*, 175–200. [CrossRef]
8. Dubbels, R.; Reiter, R.J.; Klenke, E.; Goebel, A.; Schnakenberg, E.; Ehlers, C.; Schiwara, H.W.; Schloot, W. Melatonin in edible plants identified by radioimmunoassay and by high performance liquid chromatography-mass spectrometry. *J. Pineal Res.* **1995**, *18*, 28–31. [CrossRef]
9. Hattori, A.; Migitaka, H.; Iigo, M.; Itoh, M.; Yamamoto, K.; Ohtani-Kaneko, R.; Hara, M.; Suzuki, T.; Reiter, R.J. Identification of melatonin in plants and its effects on plasma melatonin levels and binding to melatonin receptors in vertebrates. *Biochem. Mol. Biol. Intern.* **1995**, *35*, 627–634.
10. Paredes, S.D.; Korkmaz, A.; Manchester, L.C.; Tan, D.X.; Reiter, R.J. Phytomelatonin: A review. *J. Exp. Bot.* **2009**, *60*, 57–69. [CrossRef]
11. Manchester, L.C.; Tan, D.X.; Reiter, R.J.; Park, W.; Monis, K.; Qi, W. High levels of melatonin in the seeds of edible plants: Possible function in germ tissue protection. *Life Sci.* **2000**, *67*, 3023–3029. [CrossRef]
12. Tan, D.X.; Hardeland, R.; Manchester, L.C.; Korkmaz, A.; Ma, S.; Rosales-Corral, S.; Reiter, R.J. Functional roles of melatonin in plants, and perspectives in nutritional and agricultural sciences. *J. Exp. Bot.* **2012**, *63*, 577–597. [CrossRef]
13. Sturtz, M.; Cerezo, A.B.; Cantos-Villar, E.; Garcia-Parrilla, M.C. Determination of the melatonin content of different varieties of tomatoes (*Lycopersicon esculentum*) and strawberries (*Fragaria ananassa*). *Food Chem.* **2011**, *127*, 1329–1334. [CrossRef] [PubMed]
14. Garcia-Moreno, H.; Calvo, J.R.; Maldonado, M.D. High levels of melatonin generated during the brewing process. *J. Pineal Res.* **2012**, *55*, 26–30. [CrossRef] [PubMed]
15. Mercolini, L.; Saracino, M.A.; Bugamelli, F.; Ferranti, A.; Malaguti, M.; Hrelia, S.; Raggi, M.A. HPLC-F analysis of melatonin and resveratrol isomers in wine using an SPE procedure. *J. Sep. Sci.* **2008**, *31*, 1007–1014. [CrossRef]
16. Iriti, M.; Varoni, E.M.; Vitalini, S. Melatonin in traditional mediterranean diets. *J. Pineal Res.* **2010**, *49*, 101–105. [CrossRef] [PubMed]

17. Stege, P.W.; Sombra, L.L.; Messina, G.; Martinez, L.D.; Silva, M.D. Determination of melatonin in wine and plant extracts by capillary electrochromatography with immobilized carboxylic multi-walled carbon nanotubes as stationary phase. *Electrophoresis* **2010**, *31*, 2242–2248. [CrossRef]
18. Vitalini, S.; Gardana, C.; Zanzotto, A.; Fico, G.; Faoro, F.; Simonetti, P.; Iriti, M. From vineyard to glass: Agrochemicals enhance the melatonin and total polyphenol contents and antiradical activity of red wines. *J. Pineal Res.* **2011**, *51*, 278–285. [CrossRef]
19. Bonnefont-Rousselot, D.; Collin, F. Melatonin: Action as antioxidant and potential applications in human disease and aging. *Toxicol* **2010**, *278*, 55–67. [CrossRef] [PubMed]
20. Iriti, M.; Vitalini, S. Health-promoting effects of traditional mediterranean diets—A review. *Pol. J. Food Nutr. Sci.* **2012**, *62*, 71–76. [CrossRef]
21. Rodriguez-Naranjo, M.I.; Gil-Izquierdo, A.; Troncoso, A.M.; Cantos-Villar, E.; Garcia-Parrilla, M.C. Melatonin is synthesised by yeast during alcoholic fermentation in wines. *Food Chem.* **2011**, *126*, 1608–1613. [CrossRef] [PubMed]
22. Mena, P.; Gil-Izquirdo, A.; Moreno, D.A.; Martí, N.; García-Viguera, C. Melatonin is detected in pomegranate wines. Influence of variety and winemaking stage. *LWT-Food Sci. Technol.* **2012**, *47*, 13–18. [CrossRef]
23. Fernandez-Pachon, M.S.; Medina, S.; Herrero-Martin, G.; Cerrillo, I.; Berna, G.; Escudero-Lopez, B.; Ferreres, F.; Martin, F.; Garcia-Parrilla, M.C.; Gil-Izquierdo, A. Alcoholic fermentation induces melatonin synthesis in orange juice. *J. Pineal Res.* **2014**, *56*, 31–38. [CrossRef] [PubMed]
24. Gardana, C.; Iriti, M.; Stuknyte, M.; De Noni, I.; Simonetti, P. "Melatonin isomer" in wine is not an isomer of the melatonin but tryptophan-ethylester. *J. Pineal Res* **2014**, *57*, 435–441. [CrossRef] [PubMed]
25. Iriti, M.; Vigentini, I. Tryptophan-ethylester, the false (unveiled) melatonin isomer in red wine. *Int. J. Tryptophan Res.* **2015**, *8*, 1–3. [CrossRef] [PubMed]
26. Vigentini, I.; Gardana, C.; Fracassetti, D.; Gabrielli, M.; Foschino, R.; Simonetti, P.; Tirelli, A.; Iriti, M. Yeast contribution to melatonin, melatonin isomers and tryptophan ethyl ester during alcoholic fermentation of grape musts. *J. Pineal Res.* **2015**, *58*, 388–396. [CrossRef] [PubMed]
27. Fernández-Cruz, E.; Álvarez-Fernández, M.A.; Valero, E.; Troncoso, A.M.; García-Parrilla, M.C. Melatonin and derived L-tryptophan metabolites produced during alcoholic fermentation by different wine yeast strains. *Food Chem.* **2017**, *217*, 431–437. [CrossRef]
28. Mercolini, L.; Mandrioli, R.; Raggi, M.A. Content of melatonin and other antioxidants in grape-related foodstuff: Measurement using MEPS-HPLC-F method. *J. Pineal Res.* **2012**, *53*, 21–28. [CrossRef]
29. El Moussaoui, N.; Bendriss, A. Analysis of melatonin by high performance liquid chromatography after Solid-Phase Extraction (SPE/HPLC-FD). *Int. J. Eng. Res. Technol.* **2015**, *4*, 988–993.
30. Muñiz-Calvo, S.; Guillamón, J.M.; Domínguez, I.; Doménech-Carbó, A. Detecting and monitoring the production of melatonin and other related indole compounds in different *Saccharomyces* strain by solid-state electrochemical techniques. *Food Anal. Methods* **2017**, *10*, 1408–1418. [CrossRef]
31. Ribéreau-Gayon, P.; Dubourdieu, D.; Donèche, B.; Lonvaud-Funel, A. The microbiology of wine and vinifications. In *Handbook of Enology*; Ribéreau-Gayon, P., Ed.; Wiley: Hoboken, NJ, USA, 2006.
32. Magnusson, B.; Örnemark, U. Eurachem Guide: The Fitness for Purpose of Analytical Methods—A Laboratory Guide to Method Validation and Related Topics. Available online: https://www.eurachem.org/images/stories/Guides/pdf/MV_guide_2nd_ed_EN.pdf (accessed on 14 March 2018).
33. Vitalini, S.; Gardana, C.; Simonetti, P.; Fico, G.; Iriti, M. Melatonin, melatonin isomers and stilbenes in Italian traditional grape products and their antioxidant capacity. *J. Pineal Res.* **2013**, *54*, 322–333. [CrossRef] [PubMed]
34. Ramakrishna, A.; Giridhar, O.; Sankar, K.U.; Ravishankar, G.A. Melatonin and serotonin profiles in beans of Coffea species. *J. Pineal Res.* **2012**, *52*, 470–476. [CrossRef] [PubMed]
35. 2002/657/EC: Commission Decision. Official Journal of the European Communities L221, 17.8.2002. Available online: https://eur-lex.europa.eu/legal-content/EN/TXT/PDF/?uri=OJ:L:2002:221:FULL&from=HR (accessed on 14 March 2019).
36. Amerin, M.A.; Ough, C.S. *Alcohols, in Methods for Wine and Must Analysis*; Amerine, M.A., Ough, C.S., Eds.; John Wiley and Sons: New York, NY, USA, 1980.
37. Huang, X.; Mazza, G. Simultaneous analysis of serotonin, melatonin, piceid and resveratrol in fruits using liquid chromatography tandem mass spectrometry. *J. Chromatogr. A* **2011**, *1218*, 3890–3899. [CrossRef] [PubMed]

38. Gomez, F.J.V.; Raba, J.; Cerutti, S.; Silva, M.F. Monitoring melatonin and its isomer in *Vitis vinifera* cv. Malbec by UHPLC-MS/MS from grape to bottle. *J. Pineal Res.* **2012**, *52*, 349–355. [CrossRef] [PubMed]
39. Kocadağli, T.; Yılmaz, C.; Gökmen, V. Determination of melatonin and its isomer in foods by liquid chromatography tandem mass spectrometry. *Food Chem.* **2014**, *153*, 151–156. [CrossRef]
40. Spadoni, G.; Diamantini, G.; Bedini, A.; Tarzia, G.; Vacondio, F.; Silva, C.; Rivara, M.; Mor, M.; Plazzi, P.V.; Zusso, M.; et al. Synthesis, antioxidant activity and structure-activity relationships for a new series of 2-(acylaminoethyl) indoles with melatonin-like cytoprotective activity. *J. Pineal Res.* **2006**, *40*, 259–269. [CrossRef]

foods

MDPI

Article

Accumulation γ-Aminobutyric Acid and Biogenic Amines in a Traditional Raw Milk Ewe's Cheese

Rosanna Tofalo *, Giorgia Perpetuini, Noemi Battistelli, Alessia Pepe, Andrea Ianni, Giuseppe Martino and Giovanna Suzzi

Faculty of BioScience and Technology for Food, Agriculture and Environment, University of Teramo, 64100 Teramo, Italy

* Correspondence: rtofalo@unite.it; Tel.: +39-0861-266943

Received: 11 August 2019; Accepted: 5 September 2019; Published: 10 September 2019

Abstract: The influence of calf (R1), kid (R2) and pig (R3) rennets on microbiota, biogenic amines (BAs) and γ-aminobutyric acid (GABA) accumulation in raw milk ewe's cheeses was evaluated. Cheeses were investigated at different ripening times for their microbial composition, free amino acids (FAAs), BAs and GABA content. Moreover, the expression of tyrosine (*tdc*) and histidine (*hdc*) decarboxylases genes was evaluated by quantitative Real Time–Polymerase Chain Reaction (qRT-PCR). Microbial counts showed similar values in all samples. Pig rennet were cheeses were characterized by higher proteolysis and the highest values of BAs. The BAs detected were putrescine, cadaverine and tyramine, while histamine was absent. qRT-PCR confirmed this data, in fact *hdc* gene was not upregulated, while *tdc* gene expression increased over time in agreement with the increasing content of tyramine and the highest fold changes were detected in R3 cheeses. GABA showed the highest concentration in R2 cheeses reaching a value of 672 mg/kg. These results showed that the accumulation of BAs and GABA in Pecorino di Farindola is influenced by ripening time and type of coagulant. Further studies are required to develop starter cultures to reduce BAs content and improve health characteristics of raw milk ewe's cheeses.

Keywords: γ-aminobutyric acid; biogenic amines; raw milk ewe's cheese; Pecorino di Farindola; histidine decarboxylase (*hdc*) gene; tyrosine decarboxylase (*tdc*) gene

---

1. Introduction

Pecorino di Farindola is an artisanal cheese of the Abruzzo region (Italy) produced following traditional practices. Farindola is a town located in National Park of Gran Sasso, Italy, at an altitude of 530 m (1740 ft). This cheese is produced only in this geographical area and has a soft texture with a thin yellow rind or can appear with a harder texture and intense/piquant flavour as the ripening time increases. It is exclusively produced with raw ewes' milk and pig rennet without the use of natural or commercial starter cultures [1–3].

Lamb and kid rennet are the main coagulants used in Mediterranean countries for the production of Protected Denomination Origin (PDO) ovine and goat cheeses, like Pecorino Romano, Fiore Sardo and Canestrato Pugliese in Italy and Feta cheese in Greece [4,5]. Di Giacomo et al. [6] reported that pig rennet was already used by the ancient Romans to produce a "cheese of Vestini" (an ancient tribe of Abruzzo). Pig rennet is obtained from stomach mucus membrane and after an incubation of 2–3 days in salt, it is mixed with white vinegar, white wine and chili pepper and stored for 3–4 months. Finally, it is filtered for 5–6 days and only at this point it is ready to use [7–9]. Pig pepsin is unstable above pH 6.0 [9]. Its clotting activity strongly depends on pH, in fact coagulation does not occur above pH 6.7 [10]. Previous studies evaluated the influence of pig rennet in the manufacture of Pecorino di Farindola in terms of physico chemical properties, microbiota, proteolysis, volatile molecule profiles and other characteristics [3,9]. The studies showed that the use of pig makes it possible to distinguish the traditional variant from cheeses made with other coagulants.

Pecorino di Farindola cheese production is not standardized and autochthonous lactic acid bacteria (LAB) are the main responsible of the definition of final product characteristics [1]. On the other hand, non-starter lactic acid bacteria (LAB) deriving from raw milk or from the dairy environment play an important action during ripening in terms of sensory characteristics and safety issues, such as biogenic amines (BAs) accumulation [11]. Type of cheese, ripening time, manufacturing process and microorganisms highly influence the BAs content [12] and for this it can be extremely variable. BAs were found in Pecorino di Farindola cheeses examined by Schirone et al., [1]. Their total content ranged from 209.0 to 2393.0 mg/kg cheese and tyramine was the main BA detected.

BAs, mainly histamine, can lead to intoxications and adverse reactions to human health, especially after the ingestion of food products rich in BAs content [13,14]. The "cheese syndrome" caused by tyramine is rather a side effect of monoamino oxidase (MAO) drugs than a food safety issue. The consumption of cheese containing tyramine is unlikely to cause health problems in healthy individuals. EFSA report showed the absence of negative effects on health for healthy individuals after the exposure to following BA levels in food: (a) 50 mg histamine; (b) 600 mg tyramine for individuals not taking monoamino oxidase inhibitor (MAOI) drugs, but 50 mg for those using third generation MAOI drugs or 6 mg for those taking classical MAOI drugs. For putrescine and cadaverine not enough information is available [15].The intoxications related to BAs consumption can get worse in association with alcohol, other amines and monoamine and diamine oxidase-inhibiting drugs, resulting in serious problems for human health [14]. Diamine oxidase inhibitor (DAOI) drugs can also be responsible of histamine related symptomatology. Monoamine oxidase (MAO), diamine oxidase (DAO) and polyamine oxidase (PAO) are enzymes naturally present in the organism and responsible of BAs detoxification through acetylation and oxidation [16,17].

Moreover, in cheeses, there are also substances present without a defined nutritional function which could have a beneficial impact on human health and, among these compounds, γ-aminobutyric acid (GABA) has been reported with numerous positive effects on animal and human metabolic disorders [18]. This compound is synthesized by glutamate decarboxylase (GAD) catalysing the decarboxylation of L-glutamate to GABA [19]. Some studies revealed the ability of this to decrease arterial pressure and to reduce blood pressure in hypertensive patients [20–22]. Therefore, several GABA-enriched food products have been manufactured, such as GABA-enriched green tea [23], rice germ [24], tempeh and fermented beverages [25]. Dairy products fermented with GABA-producing LAB have also been studied and found to have physiological effects [26,27]. In fact, native caseins contain a high proportion of L-glutamate that can be released during milk fermentation and proteolysis.

Because of its positive impact on health, it gained the attention of the food and pharmaceutical industries. GABA is a non-protein amino acid with multiple physiological functions produced by some yeasts and bacteria. Recently, the development of functional GABA-enriched foods, such as cheese, have been reported [26,27]. In this study the role of pig rennet in the safety hazards and bioactive compounds production in Pecorino di Farindola was evaluated. In particular, cheeses were produced with three different coagulants (pig, lamb and kid) and obtained cheeses were compared in terms of microbiota, BAs and GABA.

2. Materials and Methods

2.1. Cheese Manufacture

Cheese samples were made with raw ewe's milk according to the traditional protocol [23]. Calf (R1), kid (R2) (Colombo s.r.l., Sirtori, Italy) and pig (R3) rennets (Azienda Agricola Martinelli Pietropaolo, Farindola, Italy), were used to coagulate three different milk batches obtained during the milking day, as reported by Tofalo et al. [9]. Three experimental batches consisting of 24 cheeses were carried out. After pressing and dry salting, the cheeses (~2.8 kg each) were placed in a ripening chamber at 14–15 °C. Analysis were performed after 7, 15, 30, 60, 90, 180 and 270 days of ripening.

2.2. Microbial Analysis

For the microbiological analysis, serial dilutions in sodium citrate (2% *w/v*) were prepared starting from 10 g of each cheese The following microorganisms were investigated: aerobic mesophilic bacteria (AMB) on Plate Count Agar (PCA; Oxoid, Milan, Italy) at 30 °C for 2 days; mesophilic lactobacilli on MRS agar (Oxoid, Basingstoke, UK), acidified to pH 5.4 with acetic acid, at 30 °C for 2 days in anaerobic conditions using the Gas-Pack anaerobic system (AnaeroGen; Oxoid, Basingstoke, UK); lactococci on M17 (Oxoid, Basingstoke, UK), containing 1% (*w/v*) lactose (Fluka Chimica, Milan, Italy), at 30 °C for 2 days in anaerobic conditions; enterococci on Slanetz-Bartley agar (Oxoid, Basingstoke, UK) at 37 °C for 48 h; *Enterobacteriaceae* on Violet Red Bile Glucose Agar (VRBGA; Oxoid, Basingstoke, UK) at 37 °C for 24 h. Cell counts were performed in duplicate.

2.3. qPCR Analysis

Tyrosine decarboxylase (*tdc*) and histidine decarboxylase (*hdc*) genes were detected as described by Nadkarni et al. [28], Torriani et al. [29] and Fernández et al. [30]. Primer pairs and qPCR conditions are reported in Table 1. Total RNA was extracted using a MO BIO RNA Power Soil Kit (QIAgen, Milan, Italy), according to the manufacturer's instructions. The possible presence of contaminating DNA was checked by PCR and eventually the DNase treatment was repeated. One µg of total RNA was retrotranscribed using the iScript™ cDNA Synthesis Kit (Bio-Rad, Milan, Italy), according to the manufacturer's instructions. Real-time analysis was performed using an iCycler IQ realtime PCR Detection System (Bio-Rad, Milan, Italy). A reaction mixture of 25 µL containing 12.5 µL 2XIQ SYBR Green PCR Supermix™ (Bio-Rad, Milan, Italy), 0.2 µmol/L of each primer (Life Technologies-Invitrogen, Milan, Italy) was prepared. Fold changes were determined as previously described [31]. 16S rRNA was used as reference genes. Its relative stability was evaluated using NormFinder program [32]. After real-time PCR, a melting-curve analysis was performed by measuring fluorescence during heating from 50 to 95 °C at a transition rate of 0.2 °C/s to verify the presence of unspecific products or primer dimers. A single peak was obtained highlighting the specificity of the amplification. All analyses were performed in triplicate.

Table 1. Primer sequences and PCR conditions used in this study.

Primer	Sequence (5′-3′)	qPCR Conditions	References
16SF *16SR*	TCCTACGGGAGGCAGCAGT GGACTACCAGGGTATCTAATCCTGTT	95 °C for 10 min, 40 cycles at 95 °C for 15 s, 60 °C for 1 min, 72 °C for 45 s	[28]
Tyr3 *Tyr4*	CGTACACATTCAGTTGCATGGCAT ATGTCCTACTTCTTCTTCCATTTG	94 °C for 5 min, 35 cycles at 94 °C for 20 s, 58 °C for 30 s, 72 °C for 45 s	[29]
Hdc1 *Hdc2*	TTGACCGTATCTCAGTGAGTCCAT ACGGTCATACGAAACAATACCATC	95 °C for 10 min, 40 cycles at 95 °C for 15 s, 58 °C for 1 min, 72 °C for 45 s	[30]

2.4. Free amino Acids (FAAs)

Free amino acids (FAAs, expressed as mg leucine/g) were evaluated at 507 nm after reaction with Cd-ninhydrin according to Folkertsma and Fox [33]. Analyses were performed in triplicate on each sample.

2.5. Biogenic Amines Determination

Determination of BAs was performed by acid extraction and derivatization according to Eerola, et al. (1993) [34], and Moret and Conte [35], as reported by Tittarelli et al. [17]. The presence of putrescine, cadaverine, tyramine, histamine, spermidine and spermine was determined homogenizing

2 g of cheese in 20 mL of 0.1 M HCl containing 100 mg/L of 1,7-diaminoheptane (Fluka, Milano, Italy) used as internal standard. A Waters Alliance High Performance Liquid Cromatography (HPLC) system (Waters SpA, Vimodrone, Italy), equipped with a Waters 2695 separation module connected to a Waters 2996 photodiode array detector was used. Analytes were separated using a Waters Spherisorb C18 S3ODS-2 column (3 μm particle size, 150 mm × 4.6 mm I.D.). Acetonitrile (A) and ultrapure water (B) were used for the separation of BAs. The following elution gradient was applied: 57% A for 5 min; concentration was increased up to 80% linearly in 4 min, 90% A for 5 min. The flow rate was 0.8 mL/min and the column temperature was set at 30 °C ± 0.1 °C. The peaks were detected at 254 nm. The system was controlled by Waters Empower personal computer software. Identification of the BAs was based on their retention times.

2.6. GABA Determination

GABA was determined according to Kőrös, et al. [36] and Ianni et al. [37]. Analyses were performed using the High Pressure chromatographic system (Perkin-Elmer, Monza, Italy) equipped with an autosampler and a UV-VIS detector set at 210 nm with an ion-exchange column (Phenomenex Gemini C18, dimensions: 250 × 4.6 mm, particle size: 5 μm, pore size: 110 Å) (Phenomenex, Bologna, Italy). GABA was derivatized with phthaldialdehyde Reagent (Sigma-Aldrich, Milan, Italy) and injected into the HPLC system.

3. Results and Discussion

Due to the increasing awareness towards the impact of diet on human health, issues relating to food safety and quality, have a crucial role on the consumers' behaviour. Therefore, dairy industries are developing foods with improved nutritional quality. Cheese is one of the most important fermented food product. Despite it is rich in positive compounds for human health (e.g., GABA) [38], it is also associated to BAs intoxication [13]. For this reason, the European Food Safety Authority (EFSA) Panel on Biological Hazards (BIOHAZ) has put forth an opinion on risk assessment related to BAs pointing out that the actual knowledge of their toxicity is still limited, and that further research is needed [15]. In this study, the influence of different animal rennets on microbiota, BAs and GABA accumulation in Pecorino di Farindola cheese during ripening was evaluated.

3.1. Microbial Analysis

The evolution of microbial populations at different days of ripening is reported in Figure 1. In general, no significant differences were observed. After 90 days of ripening a decrease of cell counts was detected for all microbial groups, with *Enterobacteriaceae* disappearing. The highest values of *Enterobacteriaceae* were detected in R2 cheeses with cell counts of 3.69 Log CFU/g at 90 days.

Enterobacteriaceae are an indicator of the hygienic conditions in milk and cheese production. Their occurrence has been reported in some raw milk cheeses of the Mediterranean basin after 30 days of ripening [39,40]. Moreover, some studies highlighted that this microbial group can influence taste, aroma and texture, of some artisanal cheeses [41,42]. Enterococci were present with cell counts of about 6 Log CFU/mL in all samples until 90 days of ripening, while after this time a decrease of about 3 Log was observed. Their occurrence has been reported in other Pecorino cheeses [9,43]. Their presence can be due to milk contamination and to their ability to face the conditions of cheese manufacture and ripening since they are able to develop at different temperatures and are tolerant to heat and salt [44]. Moreover, they have a crucial role in cheese ripening and aging influencing aroma and flavour thanks to their proteolytic and lipolytic activities as a result of citrate metabolism [44]. On the other hand, enterococci have often been associated with clinical infections and BAs production, such as tyramine and histamine [45].

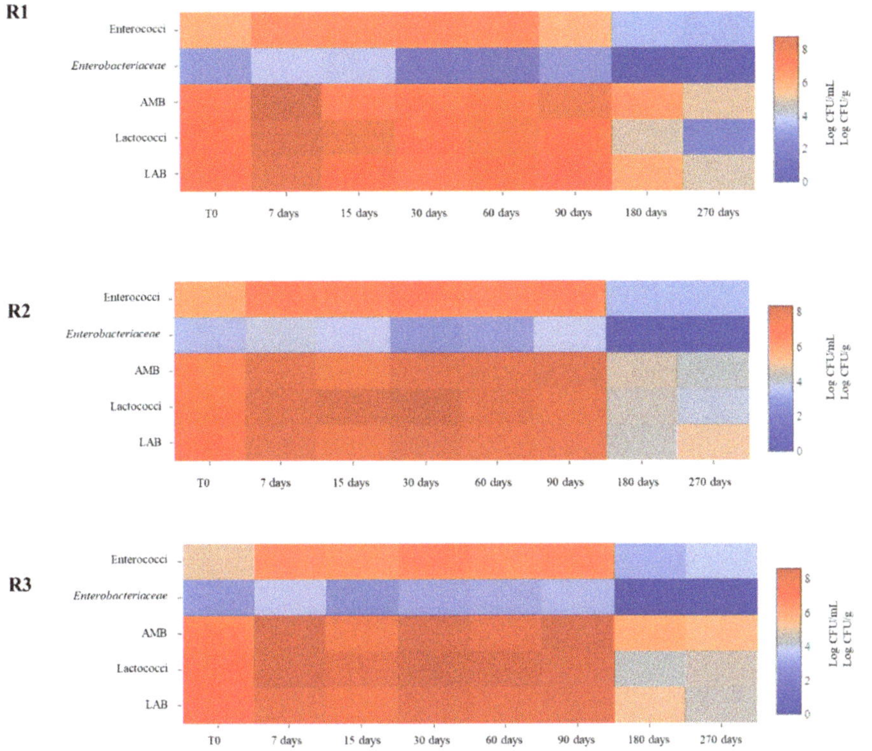

Figure 1. Heat maps depicting abundance of microbial groups characterizing the cheeses obtained with different rennets throughout ripening. Calf (**R1**), kid (**R2**) and pig (**R3**) rennets. Aerobic Mesophilic Bacteria (AMB); Lactic Acid Bacteria (LAB).

Lactococci were present in lower concentration at the end of ripening, about 4 Log CFU/mL in R2 and R3 cheeses and about 2 Log CFU/mL in R1 cheese. LAB counts throughout the ripening were similar to those observed for AMB suggesting that they were the predominant microorganisms. Similar results have been reported by Renes et al. [46]. Their counts increased during the first days of ripening reaching values of about 8 Log CFU/g in R2 and R3 cheeses after 90 days of ripening, while in R1 cheeses cell counts of about 7 Log CFU/mL were observed. LAB are the main components of the autochthonous cheese and are known to participate to the fermentation process and maturation of cheeses, producing a number of desirable substances that can improve the flavour, texture, nutritional value, shelf-life, and safety of foods [46–48]. However, some LAB species have been shown to include strains producing high amounts of BAs such as histamine and tyramine [46,49–54].

3.2. Biogenic Amines

BAs accumulation in cheese depends on several factors including ripening time, the manufacturing process, presence of decarboxylase positive microorganisms and precursors availability [14,46]. In general, during cheese ripening secondary proteolysis occurs with the accumulation of FAAs which can be decarboxylated to BAs by the microbiota [46]. Therefore, the level of protein degradation was firstly evaluated.

Figure 2 shows the evolution of FAAs during the ripening period. Obtained data revealed a certain variability of FAAs content among samples. The FAAs content increased significantly during ripening in all samples, reaching values of 233.74, 252.54 and 296.84 mg leucine/g in R1, R2 and R3 cheeses respectively after 270 days of ripening. This increase is in agreement with previous observations in Caciocavallo Pugliese, in Picante cheese, in Kashkaval cheese and in other Pecorino cheeses [9,55–58]. The greater proteolytic activity in R3 cheeses was in agreement with Tofalo et al., (2015) [9] who observed a higher proteolytic activity for pig rennet than for calf and kid ones. R3 cheeses contained the highest total concentrations of BAs (1293 mg/kg) (Figure 3). However, all the studied cheeses accumulated high total BAs contents even if with quantitative differences. It could be assumed that milk quality produced by the sheep in the Gran Sasso area is a relevant factor influencing the high BAs content of Pecorino di Farindola, probably also depending on the autochthonous microbiota [59,60].

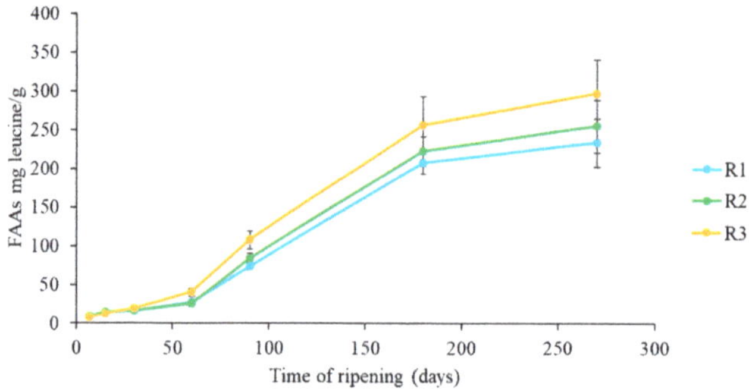

Figure 2. Evolution of free amino acids throughout ripening in Pecorino di Farindola produced with with different rennets. calf (R1), kid (R2) and pig (R3).

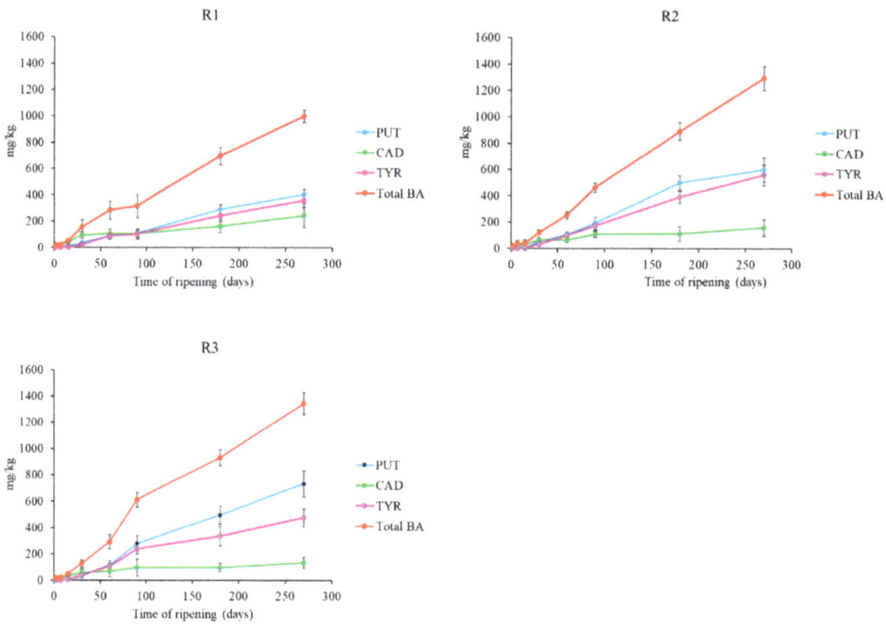

Figure 3. Biogenic amines evolution in Pecorino di Farindola manufactured with the three different rennets: calf (R1), 216 kid (R2) and pig (R3). Putrescine (PUT); Cadaverine (CAD); Tyramine (TYR); Biogenic Amine (BA).

The main BAs detected were putrescine, cadaverine and tyramine, while histamine was not detected. Putrescine showed the highest concentration with values of 400 mg/kg, 598 mg/kg and

732 mg/kg in R1, R2 and R3 cheeses, respectively after 270 days. Cadaverine presented values ranging from 241 mg/kg (R1) to 154 mg/kg (R2) at the end of ripening. These BAs are considered indicators contamination and markers of the hygiene standards of the production process. Their production mainly relies on Gram-negative bacteria, especially belonging to the families *Enterobacteriaceae* and *Pseudomonadaceae*, generally associated with spoilage [61]. LAB and staphylococci have also been reported [62]. Tyramine was present in all samples and its concentration increased over time. To our knowledge, Enterococci play a predominant role in the formation of tyramine [63]. The high BAs content detected in cheeses could be related to the productions practices associated to Pecorino di Farindola. In fact, it is produced starting from raw milk from sheep fed in a limited mountain area (Gran Sasso) and pig rennet as coagulant. The combination of these two factors could be a cause of high BAs content in the cheese. In fact, it has been already proved that there is a relationship between alpine pastures and milk quality during grazing [64].

Tyramine presents several negative effects on human health such as headaches, migraine, neurological disorders, nausea, vomiting, respiratory disorders, hypertension [13,65]. However, even though histamine was considered as the most toxic BA for a long time, recently Linares et al. [66] revealed that tyramine is even more toxic than histamine reporting the cytotoxicity threshold detected for histamine (441 mg/kg) and for tyramine (302 mg/kg). Amino acid decarboxylase activity is strain dependent rather than species specific, thus suggesting the occurrence of horizontal gene transfer events as part of a mechanism of survival and adaptation to specific environments [67]. Wüthrich et al. [68] sequenced the histamine positive strain FAM21731 of *L. parabuchneri* showing that *hdc* gene cluster was located in a genomic island, transferred within this species. This species has been frequently reported in dairy products and is responsible for the accumulation of histamine in many types of cheeses [49–51]. Its occurrence in milk is probably related to a contamination focus at farm level, because its capacity to adhere to stainless steel [50,52,53]. Moreover, *L. parabuchneri* strains are able to develop and to produce histamine also at refrigeration temperatures, suggesting that when *L. parabuchneri* is present, refrigeration can only delay but not prevent the accumulation of histamine in cheese [54]. For this reason, aminogenic strains may be found within the contaminant species but also as part of the spontaneous fermentative microbiota. Therefore, the expression of *tdc* and *hdc* genes was checked by qRT-PCR. In all samples *hdc* gene was not upregulated confirming the absence of histamine in cheeses, whereas *tdc* expression increased over time in agreement with the increasing content of tyramine (Figure 4). The strongest increase in *tdc* expression was found in R3 cheese after 270 days of ripening (a 93-fold increase), whereas the other cheeses showed 61- (R1) and 82- (R2) fold increases of the *tdc* gene.

This development was expected, because in LAB, BAs formation provides metabolic energy and/or acid resistance during a long ripening [69,70]. In general, a positive correlation between *tdc* expression and tyramine production was observed (Figure 4). This evidence is in agreement with other studies and suggests that the expression of this gene can be used to predict tyramine accumulation [13,17,43]. The high content of tyramine could be related with the high content of putrescine. In fact, agmatine deiminase pathway genes, involved in putrescine production, are linked to the tyrosine decarboxylation operon in LAB [71]. The agmatine deiminase and the tyrosine decarboxylase pathways, appear to be widespread throughout several species of LAB and are often simultaneously present [72]. Moreover, the formation of the different BAs could be also influenced by the microbial contamination in the processed milk.

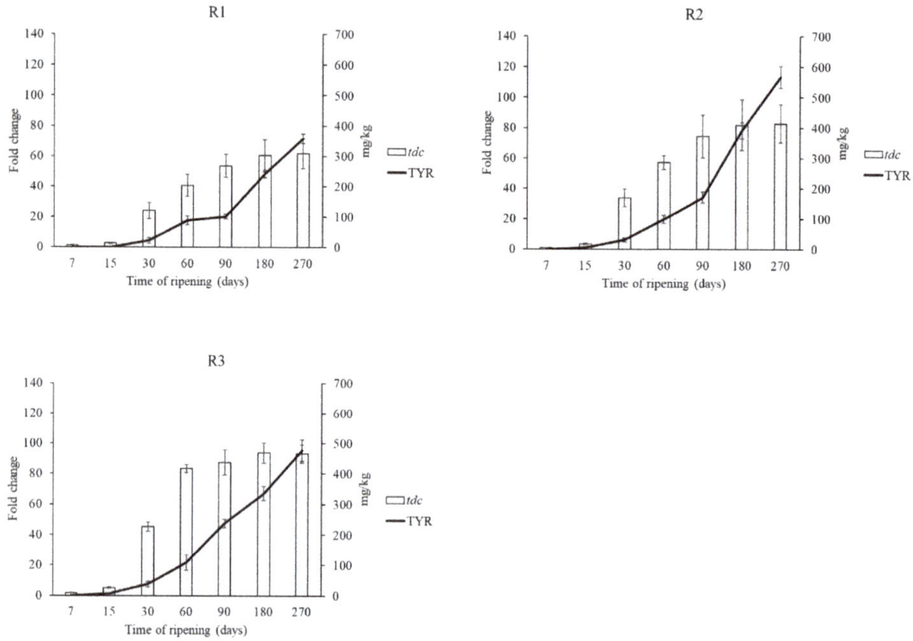

Figure 4. Tyramine content and relative transcript levels of *tdc* gene during ripening. Transcript levels are expressed as x-fold increase in comparison with the expression at T0. Three biological replicates were performed. Calf (R1), kid (R2) and pig (R3).

3.3. *γ-Aminobutyric Acid*

The GABA content determined in this study is reported in Figure 5. The highest concentrations of GABA were found in R2 and R3 cheeses, with values of 672 and 554 mg/kg, respectively.

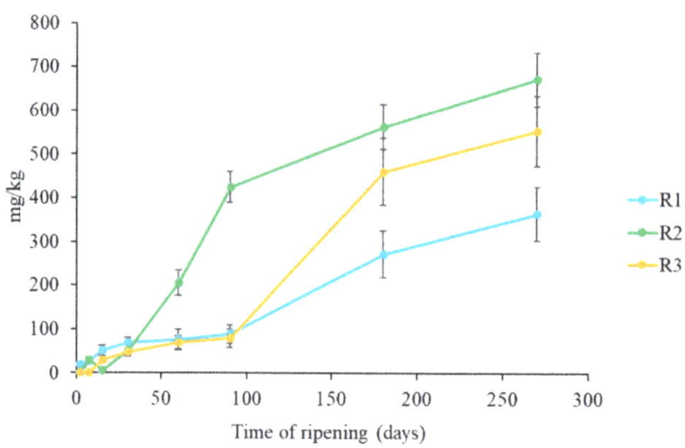

Figure 5. Evolution of GABA content in Pecorino di Farindola made with the three different rennets: calf (R1), kid (R2) and pig (R3).

These results are in agreement with the proteolysis outcome, in fact in R1 cheeses a slower proteolysis was observed. In general, GABA production is associated to different LAB species in a strain dependent way [73]. Siragusa et al. [73] studied 22 Italian cheese varieties for their GABA concentrations that varied from 0.26 to 391 mg/kg. The highest values of GABA were found in five varieties of Pecorino cheeses, but especially in Pecorino di Filiano (391 mg/kg). Moreover, Nomura et al. [74] analysed seven commercial cheeses (Camembert, Gouda, Blue, Cream, Cheddar, Edam and Emmental) and reported GABA concentration lower that those obtained in this study. In general ewe's milk quality and coagulant seem to have some influence on GABA accumulation in cheese.

4. Conclusions

Traditional cheeses like Pecorino di Farindola, maintain high diversity in cheese-making practices as well as in autochthonous cheese microbial communities. The combination of hygienic quality of raw ewe's milk, the different handling and cheesemaking processes and pig rennet could be a cause of its high BAs content. However, it is expected to see a total BAs content rise in cheese as a consequence of many BAs precursors released by coagulants from a raw milk with a high proteins content. Pig rennet imparts specific features to Pecorino di Farindola cheese, probably through its specific proteolytic patterns resulting from the unique enzymatic composition of this coagulant. However, this study confirms a greater BAs formation and proteolytic activity, as suggested by FAA values, in cheese made by pig rennet than those made by calf and kid rennet. It is difficult to modify the process without denaturing the organoleptic and sensorial characteristics of this traditional cheese. A possibility could be the pasteurization of raw milk, but this procedure denatures milk enzymes and reduces the levels of milk natural microbiota. Moreover, the pasteurized cheeses are negatively perceived by consumers compared with the sensory characteristics of cheese made with raw milk. A possible strategy for reducing BAs accumulation, increasing the safety and maintaining the sensorial characteristics of traditional cheese could be the selection of autochthonous amine-negative and amine-oxidizing LAB. These strains could be used as a starter or an adjunct/attenuated starter. In addition, Pecorino di Farindola contains high amounts of GABA which can be correlated with the use of ewe's milk, time of ripening and type of coagulant. This cheese together with other Italian "Pecorino" cheeses could be a source of microorganisms able to synthesize GABA for the production of dairy products with functional and probiotic properties.

Author Contributions: Conceptualization, R.T. and G.S.; Data curation, G.P. and N.B.; Formal analysis, N.B. and A.P.; Funding acquisition, G.M.; Methodology, G.P.; Supervision, R.T.; Writing—review & editing, R.T. and A.I.

Funding: This work is part of the following projects: "Rapid methods to determine and quantify tyramine content in dairy products of Abruzzo region" supported by grant from Cassa di Risparmio di Teramo- "Cararatterizzazione e miglioramento degli indici salutistici e sicurezza alimentare delle produzioni ovine tipiche abruzzesi a marchio di origine" (IprOv) funded by the Abruzzo Region within the Rural Development Programme 2007/2013.

Acknowledgments: The authors are grateful to "Azienda Agricola Martinelli Pietropaolo" for kind cooperation.

Conflicts of Interest: The authors declare no conflict of interest.

References

1. Schirone, M.; Tofalo, R.; Mazzone, G.; Corsetti, A.; Suzzi, G. Biogenic amine content and microbiological profile of Pecorino di Farindola cheese. *Food Microbiol.* **2011**, *28*, 128–136. [CrossRef] [PubMed]
2. Di Giacomo, F.; Casolani, N.; Signore, A. Cheese making using pig rennet and calf rennet: Microorganisms and volatile compounds in Farindola ewe cheese. *Ital. J. Food Sci.* **2014**, *26*, 2.
3. Suzzi, G.; Sacchetti, G.; Patrignani, F.; Corsetti, A.; Tofalo, R.; Schirone, M.; Fasoli, G.; Gardini, F.; Perpetuini, G.; Lanciotti, R. Influence of pig rennet on fatty acid composition, volatile molecule profile, texture and sensory properties of Pecorino di Farindola cheese. *J. Sci. Food Agric.* **2015**, *95*, 2252–2263. [CrossRef] [PubMed]
4. Addis, M.; Piredda, G.; Pirisi, A. The use of lamb rennet paste in traditional sheep milk cheese production. *Small Rumin. Res.* **2008**, *79*, 2–10. [CrossRef]
5. Addis, M.; Pirisi, A.; Di Salvo, R.; Podda, F.; Piredda, G. The influence of the enzymatic composition of lamb rennet paste on some properties of experimentally produced PDO Fiore Sardo cheese. *Int. Dairy J.* **2005**, *15*, 1271–1278. [CrossRef]
6. Di Giacomo, F.; Del Signore, A.; Giaccio, M. Pig rennet in making Farindola ewe cheese. *Prog. Nutr.* **2013**, *15*, 226–238.

7. Awad, S.; Lüthi-Peng, Q.-Q.; Puhan, Z. Proteolytic activities of chymosin and porcine pepsin on buffalo, cow, and goat whole and β-casein fractions. *J. Agric. Food Chem.* **1998**, *46*, 4997–5007. [CrossRef]
8. Boudjellab, N.; Grosclaude, J.; Zhao, X.; Collin, J.C. Development of an inhibition enzyme-linked immunosorbent assay for the detection of residual porcine pepsin in a soft cheese sample. *J. Agric. Food Chem.* **1998**, *46*, 4030–4033. [CrossRef]
9. Tofalo, R.; Schirone, M.; Fasoli, G.; Perpetuini, G.; Patrignani, F.; Manetta, A.C.; Lanciotti, R.; Corsetti, A.; Martino, G.; Suzzi, G. Influence of pig rennet on proteolysis, organic acids content and microbiota of Pecorino di Farindola, a traditional Italian ewe's raw milk cheese. *Food Chem.* **2015**, *175*, 121–127. [CrossRef]
10. Dalgleish, D.G. The enzymatic coagulation of milk. In *Cheese: Chem, Phys and Microbiol*; Springer: Boston, MA, USA, 1993; pp. 69–100.
11. Barbieri, F.; Montanari, C.; Gardini, F.; Tabanelli, G. Biogenic amine production by lactic acid bacteria: A Review. *Foods* **2019**, *8*, 17. [CrossRef]
12. Ordoñez, A.I.; Ibanez, F.C.; Torre, P.; Barcina, Y. Characterization of the casein hydrolysis of Idiazábal cheese manufactured from ovine milk. *J. Dairy Sci.* **1998**, *81*, 2089–2095. [CrossRef]
13. Ladero, V.; Martinez, N.; Martín, M.C.; Fernández, M.; Alvarez, M.A. qPCR for quantitative detection of tyramine-producing bacteria in dairy products. *Food Res. Int.* **2010**, *43*, 289–295. [CrossRef]
14. Tofalo, R.; Perpetuini, G.; Schirone, M.; Suzzi, G. Biogenic Amines: toxicology and health effect. *Encyclopedia Food Health* **2016**, 424–429. [CrossRef]
15. EFSA Panel on Biological Hazards (BIOHAZ). Scientific opinion on risk based control of biogenic amine formation in fermented foods. *EFSA J.* **2011**, *9*, 2393. [CrossRef]
16. Ruiz-Capillas, C. Biogenic amine content in Spanish retail market meat products treated with protective atmosphere and high pressure. *Eur. Food Res. Technol.* **2004**, *218*, 237–241. [CrossRef]
17. Tittarelli, F.; Perpetuini, G.; Di Gianvito, P.; Tofalo, R. Biogenic amines producing and degrading bacteria: A snapshot from raw ewes' cheese. *LWT* **2019**, *101*, 1–9. [CrossRef]
18. Diana, M.; Quilez, J.; Rafecas, M. Gamma-aminobutyric acid as a bioactive compound in foods: A review. *J. Funct. Foods* **2014**, *10*, 407–420. [CrossRef]
19. Tajabadi, N.; Baradaran, A.; Ebrahimpour, A.; Rahim, R.A.; Bakar, F.A.; Manap, M.Y.A.; Mohammed, A.S.; Saari, N. Overexpression and optimization of glutamate decarboxylase in *Lactobacillus plantarum* Taj-Apis362 for high gamma-aminobutyric acid production. *Microb. Biotechnol.* **2015**, *8*, 623–632. [CrossRef]
20. Hayakawa, K.; Kimura, M.; Kasaha, K.; Matsumoto, K.; Sansawa, H.; Yamori, Y. Effect of a γ-aminobutyric acid-enriched dairy product on the blood pressure of spontaneously hypertensive and normotensive Wistar–Kyoto rats. *Br. J. Nutr.* **2004**, *92*, 411–417. [CrossRef]
21. Kajimoto, O.; Hirata, H.; Nakagawa, S.; Kajimoto, Y.; Hayakawa, K.; Kimura, M. Hypotensive Effect of Fermented milk containing gamma-aminobutyric acid (GABA) in subjects with high normal blood pressure. *Nippon. Shokuhin Kagaku Kogaku Kaishi* **2004**, *51*, 79–86. [CrossRef]
22. Inoue, K.; Shirai, T.; Ochiai, H.; Kasao, M.; Hayakawa, K.; Kimura, M.; Sansawa, H. Blood-pressure-lowering effect of a novel fermented milk containing γ-aminobutyric acid (GABA) in mild hypertensives. *Eur. J. Clin. Nutr.* **2003**, *57*, 490–495. [CrossRef] [PubMed]
23. Ohmori, M. Effect of anaerobically treated tea on blood pressure of spontaneous hypertensive rats. *Nippon Nogeik. Kaishi* **1987**, *61*, 1449–1451. [CrossRef]
24. Leenhouts, K.; Burghoorn, J.; Brands, J.R.; Sanders, J.W.; Venema, G.; Kok, J. A chloride-inducible acid resistance mechanism in Lactococcus lactis and its regulation. *Mol. Microbiol.* **1998**, *27*, 299–310.
25. Aoki, H.; Furuya, Y.; Endo, Y.; Fujimoto, K. Effect of γ-Aminobutyric acid-enriched Tempeh-like fermented soybean (GABA-Tempeh) on the blood pressure of spontaneously hypertensive rats. *Biosci. Biotechnol. Biochem.* **2003**, *67*, 1806–1808. [CrossRef] [PubMed]
26. Park, K.-B.; Oh, S.-H. Production of yogurt with enhanced levels of gamma-aminobutyric acid and valuable nutrients using lactic acid bacteria and germinated soybean extract. *Bioresour. Technol.* **2007**, *98*, 1675–1679. [CrossRef] [PubMed]
27. Sun, T.; Zhao, S.; Wang, H.; Cai, C.; Chen, Y.; Zhang, H. ACE-inhibitory activity and gamma-aminobutyric acid content of fermented skim milk by *Lactobacillus helveticus* isolated from Xinjiang koumiss in China. *Eur. Food Res. Technol.* **2009**, *228*, 607–612. [CrossRef]
28. Nadkarni, M.A.; Hunter, N.; Jacques, N.A.; Martin, F.E. Determination of bacterial load by real-time PCR using a broad-range (universal) probe and primers set. *Microbiology* **2002**, *148*, 257–266. [CrossRef]

29. Torriani, S.; Gatto, V.; Sembeni, S.; Tofalo, R.; Suzzi, G.; Belletti, N.; Gardini, F.; Bover-Cid, S. Rapid detection and quantification of tyrosine decarboxylase gene (*tdc*) and its expression in gram-positive bacteria associated with fermented foods using PCR-based methods. *J. Food Prot.* **2008**, *71*, 93–101. [CrossRef]
30. Fernández, M.; Del Río, B.; Linares, D.; Martín, M.; Alvarez, M. Real-Time polymerase chain reaction for quantitative detection of histamine-producing bacteria: Use in cheese production. *J. Dairy Sci.* **2006**, *89*, 3763–3769. [CrossRef]
31. Livak, K.J.; Schmittgen, T.D. Analysis of relative gene expression data using real-time quantitative PCR and the $2^{-\Delta\Delta CT}$ Method. *Methods* **2001**, *25*, 402–408. [CrossRef]
32. Andersen, C.L.; Jensen, J.L.; Ørntoft, T.F. Normalization of real-time quantitative reverse transcription-PCR Data: A model-based variance estimation approach to identify genes suited for normalization, applied to bladder and colon cancer data sets. *Cancer Res.* **2004**, *64*, 5245–5250. [CrossRef] [PubMed]
33. Folkertsma, B.; Fox, P.F. Use of the Cd-ninhydrin reagent to assess proteolysis in cheese during ripening. *J. Dairy Res.* **1992**, *59*, 217–224. [CrossRef]
34. Eerola, S.; Hinkkanen, R.; Lindfors, E.; Hirvi, T. Liquid chromatographic determination of biogenic amines in dry sausages. *J. AOAC Int.* **1993**, *76*, 575–577. [PubMed]
35. Morea, M.; Matarante, A.; Di Cagno, R.; Baruzzi, F.; Minervini, F. Contribution of autochthonous non-starter lactobacilli to proteolysis in Caciocavallo Pugliese cheese. *Int. Dairy J.* **2007**, *17*, 525–534. [CrossRef]
36. Moret, S.; Conte, L.S. High-performance liquid chromatographic evaluation of biogenic amines in foods an analysis of different methods of sample preparation in relation to food characteristics. *J. Chromatogr. A* **1996**, *729*, 363–369. [CrossRef]
37. Kőrös, A.; Varga, Z.; Molnár-Perl, I. Simultaneous analysis of amino acids and amines as their o-phthalaldehyde-ethanethiol-9-fluorenylmethyl chloroformate derivatives in cheese by high-performance liquid chromatography. *J. Chromatogr. A* **2008**, *1203*, 146–152. [CrossRef] [PubMed]
38. Ianni, A.; Di Maio, G.; Pittia, P.; Grotta, L.; Perpetuini, G.; Tofalo, R.; Cichelli, A.; Martino, G. Chemical–nutritional quality and oxidative stability of milk and dairy products obtained from Friesian cows fed with a dietary supplementation of dried grape pomace. *J. Sci. Food Agric.* **2019**, *99*, 3635–3643. [CrossRef]
39. Diana, M.; Rafecas, M.; Arco, C.; Quilez, J. Free amino acid profile of Spanish artisanal cheeses: Importance of gamma-aminobutyric acid (GABA) and ornithine content. *J. Food Compos. Anal.* **2014**, *35*, 94–100. [CrossRef]
40. Montel, M.-C.; Buchin, S.; Mallet, A.; Delbès-Paus, C.; Vuitton, D.A.; Desmasures, N.; Berthier, F. Traditional cheeses: Rich and diverse microbiota with associated benefits. *Int. J. Food Microbiol.* **2014**, *177*, 136–154. [CrossRef]
41. Coton, M.; Delbès-Paus, C.; Irlinger, F.; Desmasures, N.; Le Flèche, A.; Stahl, V.; Montel, M.-C.; Coton, E. Diversity and assessment of potential risk factors of Gram-negative isolates associated with French cheeses. *Food Microbiol.* **2012**, *29*, 88–98. [CrossRef]
42. Dahl, S.; Tavaria, F.K.; Malcata, F.X.; Malcata, F. Relationships between flavour and microbiological profiles in Serra da Estrela cheese throughout ripening. *Int. Dairy J.* **2000**, *10*, 255–262. [CrossRef]
43. Schirone, M.; Tofalo, R.; Fasoli, G.; Perpetuini, G.; Corsetti, A.; Manetta, A.C.; Ciarrocchi, A.; Suzzi, G. High content of biogenic amines in Pecorino cheeses. *Food Microbiol.* **2013**, *34*, 137–144. [CrossRef] [PubMed]
44. Morales, P.; Feliu, I.; Fernández-García, E.; Nuñez, M. Volatile compounds produced in cheese by Enterobacteriaceae strains of dairy origin. *J. Food Prot.* **2004**, *67*, 567–573. [CrossRef] [PubMed]
45. Giraffa, G. Functionality of enterococci in dairy products. *Int. J. Food Microbiol.* **2003**, *88*, 215–222. [CrossRef]
46. Suzzi, G.; Caruso, M.; Gardini, F.; Lombardi, A.; Vannini, L.; Guerzoni, M.E.; Andrighetto, C.; Lanorte, M.T. A survey of the enterococci isolated from an artisanal Italian goat's cheese (semicotto caprino). *J. Appl. Microbiol.* **2000**, *89*, 267–274. [CrossRef]
47. Renes, E.; Ladero, V.; Tornadijo, M.; Fresno, J. Production of sheep milk cheese with high γ-aminobutyric acid and ornithine concentration and with reduced biogenic amines level using autochthonous lactic acid bacteria strains. *Food Microbiol.* **2019**, *78*, 1–10. [CrossRef]
48. Rodgers, S. Novel applications of live bacteria in food services: probiotics and protective cultures. *Trends Food Sci. Technol.* **2008**, *19*, 188–197. [CrossRef]
49. Settanni, L.; Moschetti, G. Non-starter lactic acid bacteria used to improve cheese quality and provide health benefits. *Food Microbiol.* **2010**, *27*, 691–697. [CrossRef]

50. Diaz, M.; Ladero, V.; Redruello, B.; Sanchez-Llana, E.; Del Rio, B.; Fernandez, M.; Martin, M.C.; Alvarez, M.A. A PCR-DGGE method for the identification of histamine-producing bacteria in cheese. *Food Control.* **2016**, *63*, 216–223. [CrossRef]
51. Ascone, P.; Maurer, J.; Haldemann, J.; Irmler, S.; Berthoud, H.; Portmann, R.; Fröhlich-Wyder, M.-T.; Wechsler, D. Prevalence and diversity of histamine-forming *Lactobacillus parabuchneri* strains in raw milk and cheese—A case study. *Int. Dairy J.* **2017**, *70*, 26–33. [CrossRef]
52. Berthoud, H.; Wüthrich, D.; Bruggmann, R.; Wechsler, D.; Fröhlich-Wyder, M.-T.; Irmler, S. Development of new methods for the quantitative detection and typing of *Lactobacillus parabuchneri* in dairy products. *Int. Dairy J.* **2017**, *70*, 65–71. [CrossRef]
53. Diaz, M.; Del Rio, B.; Sanchez-Llana, E.; Ladero, V.; Redruello, B.; Fernández, M.; Martín, M.C.; Alvarez, M.A. Histamine-producing Lactobacillus parabuchneri strains isolated from grated cheese can form biofilms on stainless steel. *Food Microbiol.* **2016**, *59*, 85–91. [CrossRef] [PubMed]
54. Diaz, M.; Ladero, V.; Del Rio, B.; Redruello, B.; Fernández, M.; Martín, M.C.; Alvarez, M.A. Biofilm-forming capacity in biogenic amine-producing bacteria isolated from dairy products. *Front. Microbiol.* **2016**, *7*, 357. [CrossRef] [PubMed]
55. Diaz, M.; Del Rio, B.; Sanchez-Llana, E.; Ladero, V.; Redruello, B.; Fernandez, M.; Martín, M.C.; Alvarez, M.A. Lactobacillus parabuchneri produces histamine in refrigerated cheese at a temperature-dependent rate. *Int. J. Food Sci. Technol.* **2018**, *53*, 2342–2348. [CrossRef]
56. Freitas, A.; Pintado, A.E.; Pintado, M.E.; Malcata, F.; Pintado, M.M.; Malcata, F. Role of dominant microflora of Picante cheese on proteolysis and lipolysis. *Int. Dairy J.* **1999**, *9*, 593–603. [CrossRef]
57. Sulejmani, E.; Hayaloglu, A.A. Influence of curd heating on proteolysis and volatiles of Kashkaval cheese. *Food Chem.* **2016**, *211*, 160–170. [CrossRef]
58. Schirone, M.; Tofalo, R.; Perpetuini, G.; Manetta, A.C.; Di Gianvito, P.; Tittarelli, F.; Battistelli, N.; Corsetti, A.; Suzzi, G.; Martino, G. Influence of iodine feeding on microbiological and physico-chemical characteristics and biogenic amines content in a raw ewes' milk cheese. *Foods* **2018**, *7*, 108. [CrossRef]
59. Curtin, A.; McSweeney, P. Catabolism of amino acids in cheese during ripening. *Cheese Chem. Phys. Microbiol.* **2004**, *1*, 435–454.
60. Hong-Xin, J.; Mi-Ya, S.; Guang-Yu, G. Influence of *Lactobacillus casei* LC2W on the proteolysis and aroma compounds of Cheddar cheese during ripening period. *CyTA J. Food* **2015**, *13*, 1–8. [CrossRef]
61. López-Caballero, M.; Sánchez-Fernández, J.; Moral, A. Growth and metabolic activity of Shewanella putrefaciens maintained under different CO2 and O2 concentrations. *Int. J. Food Microbiol.* **2001**, *64*, 277–287. [CrossRef]
62. Arena, M.; De Nadra, M.M. Biogenic amine production by *Lactobacillus*. *J. Appl. Microbiol.* **2001**, *90*, 158–162. [CrossRef] [PubMed]
63. Pircher, A.; Bauer, F.; Paulsen, P. Formation of cadaverine, histamine, putrescine and tyramine by bacteria isolated from meat, fermented sausages and cheeses. *Eur. Food Res. Technol.* **2007**, *226*, 225–231. [CrossRef]
64. Gorlier, A.; Lonati, M.; Renna, M.; Lussiana, C.; Lombardi, G.; Battaglini, L.M. Changes in pasture and cow milk compositions during a summer transhumance in the western Ital. Alps. *J. Appl. Botany Food Qual.* **2013**, *85*, 216–223.
65. Tofalo, R.; Cocchi, S.; Suzzi, G. Polyamines and gut microbiota. *Front. Nutr.* **2019**, *6*, 16. [CrossRef] [PubMed]
66. Linares, D.M.; Del Rio, B.; Redruello, B.; Ladero, V.; Martin, M.C.; Fernandez, M.; Ruas-Madiedo, P.; Alvarez, M.A. Comparative analysis of the in vitro cytotoxicity of the dietary biogenic amines tyramine and histamine. *Food Chem.* **2016**, *197*, 658–663. [CrossRef] [PubMed]
67. Coton, E.; Coton, M. Evidence of horizontal transfer as origin of strain to strain variation of the tyramine production trait in *Lactobacillus brevis*. *Food Microbiol.* **2009**, *26*, 52–57. [CrossRef]
68. Wüthrich, D.; Berthoud, H.; Wechsler, D.; Eugster, E.; Irmler, S.; Bruggmann, R. The histidine decarboxylase gene cluster of *Lactobacillus parabuchneri* was gained by horizontal gene transfer and is mobile within the species. *Front. Microbiol.* **2017**, *8*, 218. [CrossRef] [PubMed]
69. Molenaar, D.; Bosscher, J.S.; Driessen, A.J.; Konings, W.N.; Brink, B.T. Generation of a proton motive force by histidine decarboxylation and electrogenic histidine/histamine antiport in *Lactobacillus buchneri*. *J. Bacteriol.* **1993**, *175*, 2864–2870. [CrossRef]
70. Fernández, M.; Zuñiga, M. Amino Acid Catabolic Pathways of Lactic Acid Bacteria. *Crit. Rev. Microbiol.* **2006**, *32*, 155–183. [CrossRef]

71. Lucas, P.M.; Blancato, V.S.; Claisse, O.; Magni, C.; Lolkema, J.S.; Lonvaud-Funel, A.; Blancato, V. Agmatine deiminase pathway genes in *Lactobacillus brevis* are linked to the tyrosine decarboxylation operon in a putative acid resistance locus. *Microbiology* **2007**, *153*, 2221–2230. [CrossRef]
72. Arena, M.P.; Russo, P.; Capozzi, V.; Beneduce, L.; Spano, G. Effect of abiotic stress conditions on expression of the Lactobacillus brevis IOEB 9809 tyrosine decarboxylase and agmatine deiminase genes. *Ann. Microbiol.* **2011**, *61*, 179–183. [CrossRef]
73. Siragusa, S.; De Angelis, M.; Di Cagno, R.; Rizzello, C.G.; Coda, R.; Gobbetti, M. Synthesis of γ-aminobutyric acid by lactic acid bacteria isolated from a variety of Italian cheese. *Appl. Environ. Microbiol.* **2007**, *73*, 7283–7290. [CrossRef] [PubMed]
74. Nomura, M.; Kimoto, H.; Someya, Y.; Furukawa, S.; Suzuki, I. Production of γ-aminobutyric acid by cheese starters during cheese ripening. *J. Dairy Sci.* **1998**, *81*, 1486–1491. [CrossRef]

foods

MDPI

Article

Use of Selected Lactobacilli to Increase γ-Aminobutyric Acid (GABA) Content in Sourdough Bread Enriched with Amaranth Flour

Manuel Venturi [1], Viola Galli [2,*], Niccolò Pini [2], Simona Guerrini [1] and Lisa Granchi [2]

[1] FoodMicroTeam s.r.l., Via di Santo Spirito, 14, 50125 Florence, Italy; manuel@foodmicroteam.it (M.V.); simona@foodmicroteam.it (S.G.)

[2] Department of Agriculture, Food, Environment and Forestry (DAGRI), University of Florence, Piazzale delle Cascine, 18, 50144 Florence, Italy; niccolo.pini@unifi.it (N.P.); lisa.granchi@unifi.it (L.G.)

* Correspondence: viola.galli@unifi.it; Tel.: +39-055-2755917

Received: 14 May 2019; Accepted: 13 June 2019; Published: 18 June 2019

Abstract: γ-Aminobutyric acid (GABA) is the primary inhibitory neurotransmitter of the central nervous system and possesses various physiological functions. GABA production can be obtained thanks to lactic acid bacteria activity in different foods such as sourdoughs. Recently, breads made from blends of pseudocereals and wheat flours have attracted much attention. Amaranth is especially interesting because of its high nutritional value, having a high protein content and containing different antioxidant compounds. Therefore, this study aimed to obtain sourdough breads enriched with GABA thanks to bacterial activity and to investigate the effect of amaranth flour addition on the antioxidant and sensorial properties of bread. Eighteen lactobacilli strains were assayed for GABA production in amaranth and wheat flour liquid sourdoughs. Two strains, *Lactobacillus brevis* A7 and *Lactobacillus farciminis* A11, demonstrated high GABA producing capability; hence, they were used to prepare breads containing 20% amaranth flour. The results confirmed the capability of the two strains to increase GABA concentrations (up to 39 mg/kg) in breads. Samples with amaranth addition showed a significantly higher total phenolic content compared to the control bread (more than 15 mg GAE 100 g^{-1} dwb); sensory analysis showed that breads with amaranth were moderately acceptable. Nevertheless, their general liking evaluation was significantly lower compared to the control bread. The addition of pseudocereal to traditional wheat sourdough and selection of lactobacilli allowed the production of baked goods with enhanced GABA content and antioxidant capacity, but recipes have to be developed to increase the organoleptic acceptability of the final products.

Keywords: sourdough; γ-aminobutyric acid GABA; amaranth flour; lactobacilli; bioactive compounds

1. Introduction

Sourdough fermentation represents a traditional biotechnology able to improve and enhance the overall quality of leavened bakery goods obtained using a wide variety of flours [1]. Due to the activity of the occurring microbiota (lactic acid bacteria and yeasts) and the long fermentation times, sourdough positively influences the final products, improving flavour, texture, shelf-life, and above all the nutritional and functional features of the final product [2–4]. Particularly, the metabolic capability of lactic acid bacteria (LAB) is fundamental to obtain these distinctive characteristics. For instance, microbial activities toward cereal proteins, such as decarboxylation, deamination, transamination, and side chain modification, may lead to the production of several compounds which have an impact both on the nutritional quality and the flavour of bakery products [5–9]. In this context, the production of bioactive compounds, such as γ-aminobutyric acid (GABA), by LAB during sourdough fermentation could be particularly attractive [10,11]. γ-aminobutyric acid, a non-protein aminoacid, has several physiological functions in humans: Diuretic and hypotensive activity, a tranquillizing effect, and it acts as an inhibitory neurotransmitter in sympathetic brain function [10,12–15]. In fact, foods fortified with GABA are efficient in the regulation of sleeplessness, autonomic disorders, and depression [10,16]. GABA is produced from the irreversible α-decarboxylation of L-glutamic acid, catalysed by a specific enzyme, glutamic acid decarboxylase (GAD), found in bacteria, plants and animals [5,10,17]. Some works have already shown the presence of GAD activity in LAB [18–20]. In this regard, fermented foods represent an excellent source of dietary GABA [5]. Furthermore, as a stable part of diet worldwide, cereal-based products enriched with GABA are of particular interest. GABA enrichment of baked goods by exploiting selected LAB, belonging to various species, has been obtained using different flours [10,11,18,19,21]. Diana et al. [11] used a strain of *Lactobacillus brevis*, isolated from cheese, for the manufacture of wheat bread with a final GABA content of 24.2 mg/100 g. Strains of *Lactococcus lactis* and *Lactobacillus plantarum* isolated from cheese were used for the production of a bread GABA enriched, using a mixture of pseudocereals and leguminous flours [18]. The selection of peculiar cereals or pseudocereals, based on their nutritional potential, may be very useful to improve the technological and functional features of baked goods [22]. Indeed, the successful replacing of standard wheat flour with unconventional flours pseudocereal or legume flours has been demonstrated [22,23]. Whole pseudocereal grains such as quinoa, amaranth, and buckwheat have abundant content of different compounds, i.e., vitamins, fatty acids, flavonoids and phenolic acids, with a known positive impact on human health [24–27]. For these reasons, an increment of these cereals' consumption in the daily human diet has been recommended [24,28–30]. Among the pseudocereal group, amaranth is of particular interest because of its high nutritional value, having a high protein content, a high concentration of essential aminoacids [31], and containing different antioxidant compounds. Hence, the potential of sourdough fermentation, combined with amaranth supplement, could be a tool to develop new healthy baked products. Nevertheless, the sensory quality of the final products has to be taken into account since it is strongly influenced by the addition of amaranth flour [32]. Therefore, this study aimed to screen GABA-producing lactobacilli in order to obtain naturally GABA-enriched sourdough breads, and concurrently to investigate the effect of 20% addition of amaranth flour on the antioxidant properties of the final products. The sensory quality of the breads was also evaluated to determine the contribution of lactobacilli fermentation and amaranth flour addition on the organoleptic characteristics.

2. Materials and Methods

2.1. *Lactobacilli Strains and Culture Conditions*

Eighteen lactobacilli strains of five species (*L. brevis*, *L. farciminis*, *L. plantarum*, *L. rossiae* and *L. sanfranciscensis*), previously isolated from six Italian sourdoughs and belonging to the culture collection of the Department of Agriculture, Food, Environment and Forestry (DAGRI) of the University of Florence, were used in this study (Table 1).

Table 1. Lactobacilli strains and the isolation source.

Lactobacilli Strains	Italian Sourdough
L. brevis A7; *L. farciminis* A11	*Schiacciata* (flat, salty bread)
L. brevis B1; *L. sanfranciscensis* B3; *L. farciminis* B5; *L. farciminis* B7; *L. rossiae* B6; *L. farciminis* B8	Bread
L. plantarum C2	Tuscan bread
L. brevis Ga1; *L. rossiae* Ga11; *L. rossiae* Ga12; *L. rossiae* Ga14; *L. sanfranciscensis* Gd44; *L. rossiae* Gd63	*Lagaccio* biscuit
L. farciminis H3	Ancient grain bread
L. rossiae O1; *L. plantarum* O4	*Panettone* cake

2.2. *Liquid Sourdough Fermentations*

The lactobacilli strains, grown overnight in MR3i broth [33], were singly inoculated (9.0 log CFU/mL) into wheat and amaranth flour doughs. The sourdoughs were prepared by mixing water and flour to obtain a dough yield (DY), i.e., the percent ratio of the weight of the dough to the weight of the flour, of 333 and 667, for wheat and amaranth liquid sourdoughs, respectively. The samples were incubated for 6 h at 30 °C under stirring conditions (ca. 100 rpm). Control and acid control doughs (acidified to pH 3.5 by lactic acid addition) without bacterial inoculum were prepared and incubated under the same conditions. At the end of the fermentation time, sourdough samples were taken in order to perform analysis of GABA content and select the highest-GABA-producing lactobacilli strains.

2.3. Sourdough Fermentation and Bread Making

Six breads were prepared according to the recipes reported in Table 2. For each bread, a prefermented dough was prepared. Four doughs were inoculated with only baker's yeast (termed as PFC W, PFAC W, PFC Am and PFAC Am) and two with the inoculum of baker's yeast and the selected lactobacilli strains: *Lactobacillus brevis* A7 and *Lactobacillus farciminis* A11 (termed PFSD A7 and PFSD A11, respectively) (Table 2).

Table 2. Recipes for pre-fermented dough manufacture and bread making and related dough yield (dough weight × 100/flour weight). PF—prefermented; C—control; AC—acid control; SD—sourdough; W—wheat flour; Am—Amaranth flour; A7—*L. brevis* A7 strain; A11—*L. farciminis* A11 strain.

			Prefermented Dough			
Ingredients (% on Total Flour)	**PFC W**	**PFAC W**	**PFC Am**	**PFAC Am**	**PFSD A7**	**PFSD A11**
Wheat flour	100	100	80	80	80	80
Amaranth flour	–	–	20	20	20	20
Water	58	58	58	58	58	58
Baker's yeast	1	1	1	1	1	1
L. brevis A7 (log (CFU/g))	–	–	–	–	9	–
L. farciminis A11 (log (CFU/g))	–	–	–	–	–	9
			Bread			
Ingredients (% on total flour)	**C W**	**AC W**	**C Am**	**AC Am**	**SD-A7**	**SD-A11**
Prefermented dough	50	50	50	50	50	50
Wheat flour	100	100	80	80	80	80
Amaranth flour	–	–	20	20	20	20
Water	58	58	58	58	58	58
Dough yield	158	158	158	158	158	158

Amaranth flour was added as 20% of the total flour in four doughs; acid control doughs were acidified by lactic acid to 3.5. The prefermented doughs were fermented for 18 h at 30 °C and then added to the final mixtures. In the final dough (Table 2) the ingredients were added at the same time and mixed for 10 min in a twin arms mixer (model RS12, Bernardi, Italy). The doughs were placed in the trays at 30 °C with 88–90% relative humidity for 3 h. Samples were taken at the beginning and at the end of the leavening time. Finally, doughs were baked at 180 °C for 15 min.

2.4. Monitoring of Selected Lactobacilli Strains in Sourdough Fermentations

A Randomly Amplified Polymorphic DNA (RAPD) analysis was performed according to Venturi et al. [34] to assess the occurrence of the selected lactobacilli in liquid and firm sourdoughs and final doughs. DNA was amplified using the following primers: OPL-05 (50 ACGCAGGCA 30), designed by Seseña et al. [35], and MV1 (50 GGACGCTTCTG 30) designed by Venturi et al. [34]. The random primer MV1 was used separately, while OPL-05 was used along with the primer RD1 (5′ GCTTAAGGAGGTGATCCAGCC 3′). DNA amplification was performed as described by Reguant and Bordons [36]. Amplification products were separated (at 100 V for 2.5 h) on 1.4% (*w*/*v*) agarose gel (Lonza Group Ltd, Basel, Switzerland); containing ethidium bromide (Sigma e Aldrich, St Louis, MI, USA) and TEB buffer (1 M Tris, 10 mM EDTA, 0.9 M boric acid, pH 8.3). The resulting profiles were captured as images after UV transillumination, and compared to those previously obtained for each bacterial strain.

2.5. Determination of pH, Total Titratable Acidity, Volume Increase, and Enumeration of Cultivable Bacteria and Yeasts

Ten grams of dough sample were transferred into 90 mL of sterile physiological solution, homogenized for 2 min in a Stomacher Lab Blender 400 (Seward Ltd, Worthing, West Sussex, UK). After decimal dilutions, 100 μL of these suspensions were plated for cell enumeration using MR3i medium for the lactobacilli and MYPG for the baker's yeast using the pour plate method. Lactobacilli were counted after incubation for 48–72 h at 30 °C under anaerobic conditions. Yeasts, plated on MYPG agar containing sodium propionate (2 g/L), were counted after incubation for 48 h at 30 °C under aerobic conditions. Plate counts were performed in duplicate. The pH values were determined by a pH-meter (Metrohm Italiana Srl, Varese, Italy). Total titratable acidity (TTA) was measured on 10 g of dough samples, which were homogenized with 90 mL of distilled water for 3 min and expressed as the amount (mL) of 0.1 N NaOH to achieve a pH of 8.5. To assess sourdough increase of volume, 100 g of each dough were placed in a graduated cylinder (0.5 L). The volume of the doughs (in mL) was recorded immediately (t_0) and after 3 h of fermentation at 30 °C. The leavening was calculated using the following formula: $((V_3 - V_0)/V_0) \times 100$, where V_3 was the volume after the 3 h fermentation and V_0 was the initial volume.

2.6. Determination of Lactic Acid by HPLC

Bread samples were diluted ten times with distilled water and then filtered by Amicon® Ultra-4 Centrifugal Filters (3000 Da NMWL) (Merck Millipore) before the injection for lactic acid determination by high-performance liquid chromatography (HPLC) analysis (Varian Inc., Palo Alto, CA, USA) connected to a refractive index detector (Knauer K-2301, Knauer GmbH, Berlin, Germany) and UV detector (λ = 210). Elution was performed at 65 °C with 0.01 N H_2SO_4 eluent at a flow rate of 0.6 mL/min. Data were collected and analysed by using the Galaxie software (Varian Inc., Palo Alto, CA, USA). Quantitative analysis was carried out by a standard curve.

2.7. Total Phenols Assay by Folin–Ciocalteau Reagent

Total phenols content of bread samples was determined according to Alvarez-Jubete et al. [37], with some modifications. Bread samples were dried for 24 h at 60 °C and ground. 1.25 g were weighted and added to 25 mL methanol. Samples were vortexed and left in horizontal shaking for 24 h, then centrifuged for 10 min at 2000× *g*. The final extracts were obtained by filtering 10 mL of the supernatant through 0.22 μm PTFE syringe filters (Whatman) and they were stored at −20 °C until analysis. The reaction mixture consisted of 100 μL of methanolic bread extract, 100 μL of methanol, 100 μL of Folin–Ciocalteu reagent and 700 μL of Na_2CO_3. The samples were vortexed immediately, and the tubes were incubated in the dark for 20 min at room temperature. After incubation, all samples were centrifuged at 11,300× *g* for 3 min. The absorbance of the supernatant was then measured at 735 nm in 1 mL plastic cuvette using a spectrophotometer (Cary 50 Scan, Varian Inc., Palo Alto, CA, USA). Gallic acid was used as a standard and a calibration curve was prepared with a range of concentrations from 10–200 mg/L. The results are expressed in mg of gallic acid equivalent per 100 g of dry-weight basis (mg GAE 100 g^{-1} dwb).

2.8. Antioxidant Capacity by DPPH Assay

The free radical scavenging capacity of methanolic bread extracts was determined using the stable 2,2-diphenyl-1-picrylhydrazyl radical (DPPH). The scavenging effect was measured according to the method of Alvarez-Jubete et al. [37]. Also, 500 μL of extracts were added to 500 μL of DPPH methanolic solution (0.05 mg/mL). After vortexing, the mixture was left for 40 min at room temperature, and the absorbance of the resulting solution was read at 517 nm. The absorbance measured after 40 min was used for the calculation of the antioxidant capacity according to the following formula: DPPH radical-scavenging activity (%): ((blank absorbance − sample absorbance)/blank absorbance) × 100. Butylated hydroxytoluene (BHT) was also assayed as antioxidant references.

2.9. HPLC Determination of GABA Content

GABA content was determined on the water-soluble extracts of liquid doughs and breads. The extracts were obtained by extracting the samples with sterile distilled water (1:3 *w*/*v*), held at 4 °C for 1 h and vortexed at 15-min intervals. Samples were finally centrifuged at 14,000× *g* for 20 min. The supernatants, containing the water-soluble fraction, were used for GABA quantification. Before the injection, the samples were prepared according to Tuberoso et al. [38]. The reaction mixture consisted of 100 µL of sample extracts, 5 µL of 100 mg/L eptilammine (internal standard, IS), 200 µL of dansyl chloride solution (derivatization agent) and 0.2 M $Na_2B_4O_7{\cdot}10H_2O$ (pH 9.3) solution up to a final volume of 1000 µL. The mixture was incubated for 30 min at 40 °C in a Termoblok and centrifuged at 11,300× *g* for 10 min. The supernatant was recovered and diluted with MeOH (1:1 *v*/*v*) for HPLC analysis. Separation was obtained with a Phenomenex Gemini C18 110A column (150 4.60 mm, 3 lm; Chemtek Analitica, Anzola Emilia, Bologna, Italy) connected to fluorimetric detector (Jasco Europe, Cremella, LC, Italy) with wavelengths set at 293 nm (Ex) and 492 nm (Em) under the following conditions: Mobile phases buffer acetate/CH_3CN (pH 4.1) and acetonitrile, flow rate 0.8 mL/min, column temperature 25 °C. The quantitative analysis was performed using calibration graphs constructed according to the internal standard method.

2.10. Sensory Evaluation

Sensory evaluation of lactobacilli inoculated breads and non-acidified control breads was carried out by 46 panellists (15 male and 31 female) aged 21–65 years old. Breads were cut into pieces of 3 cm × 3 cm and randomly codified before serving. Colour, aroma, consistency, and general liking were evaluated using the hedonic 9-point scale (from "1-dislike extremely" to "9-like extremely") [39]. Panellists have also optionally chosen among 13 attributes for the taste of breads: Salty, vapid, sweet, earthy, strange taste, sour, delicate, tasty, astringent, not to eat, gummy, interesting, and persistent [24].

2.11. Statistical Analysis

Chemical and microbiological determinations, performed in duplicate, were elaborated according to *t*-Test procedures or nonparametric one-way ANOVA followed by Tukey's Test (Statistica 7.0 software package, Stat Software Inc., Tulsa, OK, USA). Differences were reported at a significance level of $p \leq 0.05$.

3. Results and Discussion

3.1. Selection of GABA-Producing Lactobacilli Strains

Eighteen lactobacilli strains were singly inoculated in liquid doughs made with either wheat or amaranth flour and incubated at 30 °C for 6 h in order to evaluate the GABA production. All the strains were able to acidify the doughs although some differences were registered. The final pH values ranged from ca 3.50–4.10 with an average decrease of 2.35 ± 0.13 in wheat sourdoughs. Due to the lower flour content (DY = 667) and a different buffering capacity, the final pH of amaranth sourdoughs resulted higher, in a range from 4.10–5.10, with an average pH decrease of 1.74 ± 0.41. The most acidifying strains, both in wheat and amaranth flour, were *L. brevis* A7, *L. farciminis* A11, *L. farciminis* B7 and *L. rossiae* Ga12. The values of pH of the non-inoculated doughs (C) were 5.96 ± 0.05 and 6.20 ± 0.11 for wheat and amaranth dough, respectively. After 6 h of fermentation GABA concentrations were determined.

Results showed a variability ($p \leq 0.05$) among the tested strains in both the flours, even if GABA concentrations were generally higher in wheat sourdoughs (Figure 1). Differences in GABA concentration in sourdoughs produced by the two flours could be due to higher glutamic acid content in wheat flour, since it is the substrate for glutamic acid decarboxylase enzyme (GAD). In agreement with the literature, the GABA producing activity was not related to the species, resulting in strain dependency [40]. In wheat sourdoughs, six lactobacilli strains displayed GABA concentrations significantly higher than the control and the acidified control, whereas in amaranth sourdoughs this trend was observed only for one strain (*L. rossiae* Gd63). Particularly, the highest ($p \leq 0.05$) GABA concentration in wheat sourdoughs was found in the dough inoculated with *L. brevis* A7 (136.62 ± 4.00 mg/kg of flour), followed by *L. plantarum* O4, *L. brevis* Ga1, *L. farciminis* H3, *L. rossiae* O1, *L. farciminis* A11. The strain *L. rossiae* Gd63 displayed the highest GABA production in amaranth sourdough fermentation (47.6 ± 14.3 mg/kg of flour), although other six lactobacilli strains (*L. plantarum* C2, *L. brevis* A7, *L. sanfranciscensis* Gd44, *L. brevis* B1, *L. rossiae* Ga14 and *L. farciminis* A11) did not show statistical differences from the GABA amounts produced by the best performing strain. Hence, based on the reported results concerning GABA production and acidification ability, *L. farciminis* A11 and *L. brevis* A7 were selected to prepare firm sourdoughs and breads.

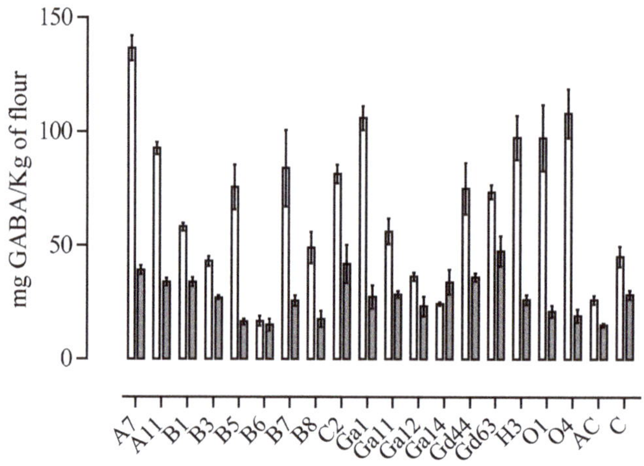

Figure 1. Production of GABA after 6 h fermentation in wheat (white bar) and amaranth (grey bar) sourdoughs obtained with 18 lactobacilli strains singly inoculated (data are expressed as mean ± coefficient of variation %). C—control; AC—acid control.

3.2. Sourdough Bread Fermentation with Selected Lactobacilli Strains

Breads were prepared with the addition of 20% of amaranth flour. This percentage was chosen because it is considered appropriate by various authors in order to maintain product structural quality [41], preserve the principal nutritional benefit of this ingredient [42], and meet sensory approval [43]. Table 3 shows the main technological and microbiological characteristics at the end of fermentation of the sourdoughs prepared with the inoculum of *L. brevis* A7 (SD-A7) and *L. farciminis* A11 (SD-A11). The occurrence of the two inoculated strains throughout the process was confirmed by molecular analysis. In particular, all the lactobacilli isolates from sourdoughs displayed solely the genotypic patterns of *L. brevis* A7 or *L. farciminis* A11.

Table 3. Values (mean ± standard deviation) of pH, Total Titratable Acidity (TTA; mL NaOH), lactic acid (g/kg), volume increase (%), and microorganism (lactobacilli and baker's yeast) concentrations (log (CFU/g)) after sourdough fermentations carried out by *L. brevis* A7 (SD-A7) and *L. farciminis* A11 (SD-A11).

Dough	Final pH	ΔpH	Final TTA	ΔTTA	Lactic acid	ΔV/V0 × 100	Lactobacilli	Yeasts
SD-A7	4.35 ± 0.03 ^{a}	0.95 ± 0.39 ^{a}	6.90 ± 0.14 ^{b}	2.80 ± 0.66 ^{a}	4.60 ± 0.63 ^{b}	105 ± 7.07 ^{a}	8.50 ± 0.72 ^{a}	7.35 ± 0.64 ^{a}
SD-A11	4.60 ± 0.18 ^{b}	0.82 ± 0.10 ^{a}	6.45 ± 0.07 ^{a}	2.95 ± 0.78 ^{a}	3.20 ± 0.56 ^{a}	105 ± 21.2 ^{a}	8.59 ± 0.73 ^{a}	7.40 ± 0.62 ^{a}

ΔV = difference between the final and the initial value; V_0—initial volume. Values in the same column with different letters (**a**,**b**) are significantly different ($p \leq 0.05$).

The 20% amaranth incorporation did not affect the considered features. At the end of leavening time, both the sourdough pH decreased of about 0.88, reaching a value of 4.35 ± 0.03 for SD-A7 and 4.60 ± 0.18 for SD-A11, statistically different from each other and showing proper acidification [44,45]. SD-A11 final total titratable acidity was lower compared to SD-A7, while the TTA increase did not point out any significant differences. Regarding organic acids production, the highest content of lactic acid was found in SD-A7 dough (4.60 ± 0.63 g/kg). Both the samples doubled their volume, indicating an adequate leavening process. Microbiological analyses did not show any significant differences in the microorganism concentrations at the end of the fermentation. *L. farciminis* A11 and *L. brevis* A7 reached a concentration of about 8.55 log CFU/g, which is a typical value in sourdough [46], and *S. cerevisiae* reached a concentration of 7.37 log CFU/g.

3.3. Phenolic Content and Antioxidant Capacity of Breads

The results of the total phenolic content, determined on the methanolic extracts of all the breads, are shown in Figure 2.

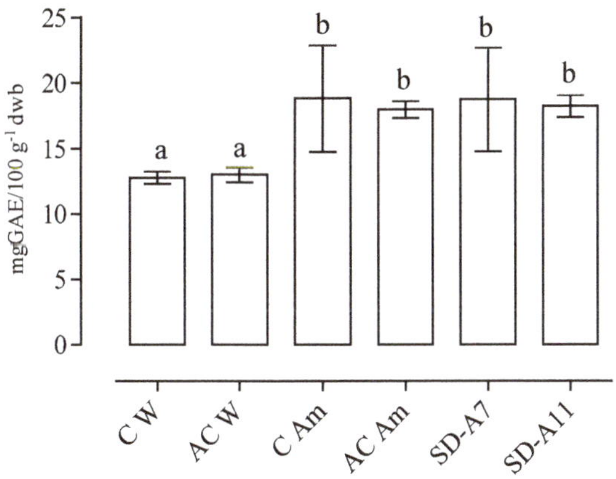

Figure 2. Total phenolic content (expressed as mg of gallic acid equivalents GAE100 g^{-1} dwb) of the breads (data are expressed as mean ± coefficient of variation %). C W—wheat control bread; AC W—wheat acid control bread; C Am—amaranth control bread; AC Am—amaranth acid control bread; SD-A7—*L. brevis* A7 bread; SD-A11—*L. farciminis* A11 bread. Different letters (**a**,**b**) indicate significant differences ($p \leq 0.05$).

Results indicated that amaranth flour addition significantly increased the phenolic content of breads. Indeed, whole pseudocereal grains are rich in this class of minor components, therefore their use as a supplement can increase the nutritional value of baked goods. Data showed that samples integrated with 20% of amaranth flour displayed a phenolic content of more than 15 mg GAE 100 g^{-1} dwb, which was significantly higher than the control breads, C W and AC W, made only with wheat flour ($p \leq 0.05$). These results are in agreement with those reported by Chlopicka et al. [24], which showed a higher content of phenolic compounds in breads enriched with amaranth flour, compared to a control bread without this pseudocereal addition.

The radical-scavenging capacity of the breads was tested by the DPPH method (Figure 3).

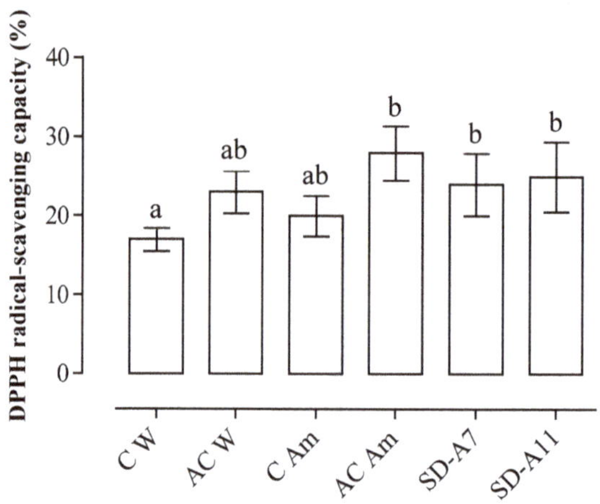

Figure 3. Radical-scavenging capacity (%) of the breads (data are expressed as mean ± coefficient of variation %). C W—wheat control bread; AC W—wheat acid control bread; C Am—amaranth control bread; AC Am—amaranth acid control bread; SD-A7—*L. brevis* A7 bread; SD-A11—*L. farciminis* A11 bread. Different letters (**a**,**b**) indicate significant differences ($p \leq 0.05$).

The antioxidant capacity of control bread made with only wheat flour result was the lowest among the samples (less than 17%), while AC Am, SD-A7 and SD-A11 showed the highest values ($p \leq 0.05$), up to 28%. According to Vollmannova et al. [47], total phenolic content was partially correlated to the antioxidant capacity; indeed, AC Am, SD-A7 and SD-A11 showed the highest values for both these parameters. Amaranth addition only slightly increased the antioxidant capacity, on the contrary, the acidification seems to have a higher role in increasing this activity. In fact, both the chemically acidified controls showed higher values than the controls. Hence, the combination of amaranth addition and acidification might be responsible of the observed phenomenon. In this context, it was demonstrated that acidification, as a consequence of microbial fermentation, can increase the levels of easily-extractable phenolic compounds and their bioavailability [4,48,49].

3.4. GABA Content of Breads

GABA concentrations in the breads prepared by lactobacilli inoculated sourdoughs and in the control breads are reported in Figure 4.

GABA content in the breads ranged from 8.0–39.0 mg/kg. Sourdough breads showed the highest concentration with values of 26.9 ± 1.53 and 39.0 ± 1.53 mg/kg for SD-A7 and SD-A11, respectively, confirming the capability of LAB to produce GABA. The lowest content was detected in the control breads with 100% of wheat flour, less than 10 mg/kg. Compared to the control wheat breads, the GABA increase of SD-A11 was of about 350%. Other studies exploited high GABA producing biotype of *Lactobacillus plantarum* and *Lactococcus lactis* in order to obtain a bread with improved nutritional features [10,18,19]. The ability of *L. brevis* strains to synthetize GABA was reported in various food matrices such as cheese [40], yoghurt [50], black raspberry juice [51], and wheat sourdough [11], while this capability was observed only in a strain of *L. farciminis* isolated from Myanmar traditional fermented fishery products with boiled rice [52]. To the best of our knowledge, this was the first time in which GABA production by a strains of *L. farciminis* was exploited in a bakery product.

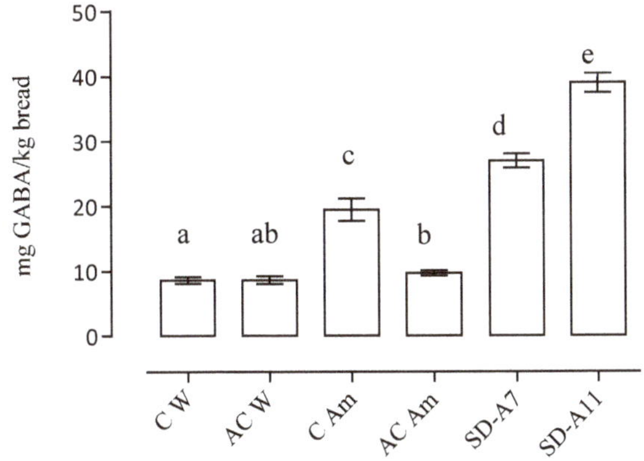

Figure 4. GABA concentration (mg/kg) in the breads (data are expressed as mean ± standard deviation). C W—wheat control bread; AC W—wheat acid control bread; C Am—amaranth control bread; AC Am—amaranth acid control bread; SD-A7—*L. brevis* A7 bread; SD-A11—*L. farciminis* A11 bread. Different letters (**a–c**) indicate significant differences ($p \leq 0.05$).

3.5. Sensory Evaluation

To evaluate the contribution of amaranth flour and lactobacilli fermentation on organoleptic characteristics of breads, a sensory evaluation was carried out with a panel of 46 members. Table 4 shows the results of a 9-point hedonic scale analysis of colour, odour, consistency, and general liking.

Table 4. Sensory evaluation of breads (mean ± standard deviation). Values were determined with a 9-point scale (extremely disliked = 1, neither like, nor dislike = 5, extremely liked = 9). C W—wheat control bread; C Am—amaranth control bread; SD-A7—*L. brevis* A7 bread; SD-A11—*L. farciminis* A11 bread.

	C W	C Am	SD-A7	SD-A11
Colour	6.7 ± 1.3 ^{a}	6.2 ± 1.2 ^{a}	6.1 ± 1.2 ^{a}	6.0 ± 1.5 ^{a}
Aroma	6.1 ± 1.5 ^{a}	5.6 ± 1.6 ^{a}	5.6 ± 1.5 ^{a}	5.3 ± 2.0 ^{a}
Consistency	6.3 ± 1.4 ^{b}	5.8 ± 1.6 ^{ab}	5.1 ± 1.6 ^{a}	5.2 ± 1.5 ^{a}
General liking	6.3 ± 1.5 ^{b}	5.3 ± 1.6 ^{a}	4.9 ± 1.6 ^{a}	4.9 ± 1.8 ^{a}

Values in the same column with different letters (**a**,**b**) are significantly different ($p \leq 0.05$).

Values of colour and aroma were not significantly different among the breads, ranging from 6–6.7 and from 5.3–6.1, hence indicating a moderate appreciation. On the contrary, the consistency of bread was negatively affected by the addition of amaranth flour; in fact, the best value was registered for the control wheat bread (6.3 ± 1.4), statistically higher ($p \leq 0.05$) compared to the inoculated amaranth breads (5.1–5.2). General liking of the 100% wheat control bread showed a significantly higher value (6.3 ± 1.5) compared to bread with amaranth replacement; however, none of the breads were negatively evaluated by the panellists. These results pointed out that substituting 20% of wheat flour with amaranth flour in bread was not advantageous for bread sensory evaluation, even if the panellists gave a moderate appreciation. In addition, other authors [42] reported that, considering the better nutritional features, consumers would choose to consume amaranth bread instead of common wheat bread, even if the taste is different. Furthermore, the spider plot in Figure 5 shows the different attributes chosen by panellists to describe breads.

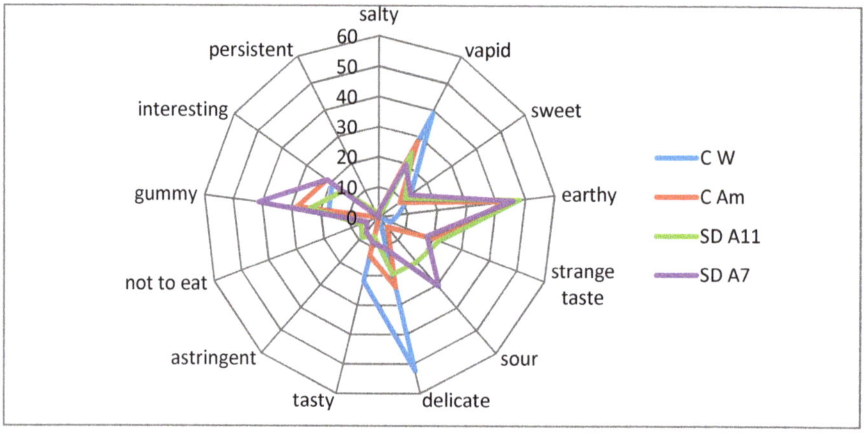

Figure 5. Spider plot of the sensory evaluation (%) of breads. Blue line: % of positive answers given to wheat control bread (C W); red line: % of positive answers given to amaranth control bread (C Am); green line: % of positive answers given to *L. farciminis* A11 sourdough bread (SD-A11); purple line: % of positive answers given to *L. brevis* A7 sourdough bread (SD-A7).

Amaranth clearly characterized bread for earthy taste (about 50% panellists for all the amaranth breads). More than 50% panellists defined the wheat control bread as delicate, even if 40% also defined breads as vapid with respect to 20–30% of the breads with amaranth. In particular, panellists declared that SD-A7 bread was sour (30%) and gummy (more than 40%). Nevertheless, about 20% of panellists described as interesting the taste of all the breads, and only a few testers (about 5%) declared SD amaranth breads "not to eat", in contrast with the 10–30% reported in other experiments [24]. Sensory analysis highlighted some differences between the two sourdough amaranth breads. SD-A7 bread taste was described as less delicate, gummier, and sourer, but also more interesting compared to SD-A11 bread. However, results showed that the 20% amaranth incorporation did not strongly reduce the breads' appreciation.

4. Conclusions

The screening of GABA-producing lactobacilli in liquid sourdoughs led to the selection of two strains belonging to *L. farciminis* and *L. brevis* species. These strains were able to increase GABA concentration (up to 350%) in breads enriched with 20% of amaranth flour, with a final content of 26.9 ± 1.53 and 39.0 ± 1.53 mg/kg respectively. The final products showed higher antioxidant activity and an increased content of total phenolic compounds compared to the wheat control bread. Sensory evaluation indicated moderate acceptability of breads with amaranth flour, mainly characterized by an earthy taste. Although organoleptic acceptability of the final products could be improved, the combination of selected lactobacilli and pseudocereal flours can be a suitable tool for the production of innovative baked goods with improved nutritional features.

Author Contributions: Conceptualization, M.V. and V.G.; methodology, M.V. and V.G.; formal analysis, M.V., V.G. and N.P.; investigation, M.V., V.G. and N.P.; data curation, M.V., V.G. and L.G.; writing—original draft preparation, M.V. and V.G; writing—review and editing, M.V., V.G., S.G. and L.G.; supervision, S.G. and L.G.

Funding: This research received no external funding.

Conflicts of Interest: The authors declare no conflict of interest.

References

1. Chavan, R.S.; Chavan, S.R. Sourdough technology—A traditional way for wholesome foods: A review. *Compr. Rev. Food Sci. Food Saf.* **2011**, *10*, 169–182. [CrossRef]
2. Arendt, E.K.; Ryan, L.A.; Dal Bello, F. Impact of sourdough on the texture of bread. *Food Microbiol.* **2007**, *24*, 165–174. [CrossRef] [PubMed]
3. Hammes, W.P.; Gänzle, M.G. Sourdough breads and related products. In *Microbiology of Fermented Foods*, 2nd ed.; Woods, B.J.B., Ed.; Blackie Academic: London, UK, 1998; pp. 199–216.
4. Gobbetti, M.; De Angelis, M.; Di Cagno, R.; Calasso, M.; Archetti, G.; Rizzello, C.G. Novel insights on the functional/nutritional features of the sourdough fermentation. *Int. J. Food Microbiol.* **2018**. [CrossRef] [PubMed]
5. Diana, M.; Quílez, J.; Rafecas, M. Gamma-aminobutyric acid as a bioactive compound in foods: A review. *J. Funct. Foods* **2014**, *10*, 407–420. [CrossRef]
6. Di Cagno, R.; De Angelis, M.; Lavermicocca, P.; De Vincenzi, M.; Giovannini, C.; Faccia, M.; Gobbetti, M. Proteolysis by sourdough lactic acid bacteria: Effects on wheat flour protein fractions and gliadin peptides involved in human cereal intolerance. *Appl. Environ. Microbiol.* **2002**, *68*, 623–633. [CrossRef]
7. Gänzle, M.G.; Loponen, J.; Gobbetti, M. Proteolysis in sourdough fermentations: Mechanisms and potential for improved bread quality. *Trends Food Sci. Technol.* **2008**, *19*, 513–521. [CrossRef]
8. Gobbetti, M.; Rizzello, C.G.; Di Cagno, R.; De Angelis, M. How the sourdough may affect the functional features of leavened baked goods. *Food Microbiol.* **2014**, *37*, 30–40. [CrossRef]
9. Paterson, A.; Piggott, J.R. Flavour in sourdough breads: A review. *Trends Food Sci. Technol.* **2006**, *17*, 557–566. [CrossRef]
10. Bhanwar, S.; Bamnia, M.; Ghosh, M.; Ganguli, A. Use of *Lactococcus lactis* to enrich sourdough bread with γ-aminobutyric acid. *Int. J. Food Sci. Nutr.* **2013**, *64*, 77–81. [CrossRef]
11. Diana, M.; Rafecas, M.; Quílez, J. Free amino acids, acrylamide and biogenic amines in gamma-aminobutyric acid enriched sourdough and commercial breads. *J. Cereal Sci.* **2014**, *60*, 639–644. [CrossRef]
12. Huang, J.; Mei, L.H.; Wu, H.; Lin, D.Q. Biosynthesis of γ-aminobutyric acid (GABA) using immobilized whole cells of *Lactobacillus brevis*. *World J. Microbiol. Biot.* **2007**, *23*, 865–871. [CrossRef]
13. Komatsuzaki, N.; Shima, J.; Kawamoto, S.; Momose, H.; Kimura, T. Production of γ-aminobutyric acid (GABA) by *Lactobacillus paracasei* isolated from traditional fermented foods. *Food Microbiol.* **2005**, *22*, 497–504. [CrossRef]
14. Komatsuzaki, N.; Tsukahara, K.; Toyoshima, H.; Suzuki, T.; Shimizu, N.; Kimura, T. Effect of soaking and gaseous treatment on GABA content in germinated brown rice. *J. Food Eng.* **2007**, *78*, 556–560. [CrossRef]
15. Su, Y.C.; Wang, J.J.; Lin, T.T.; Pan, T.M. Production of the secondary metabolites γ-aminobutyric acid and monacolin K by *Monascus*. *J. Ind. Microbiol. Biot.* **2003**, *30*, 41–46. [CrossRef] [PubMed]
16. Okada, T.; Sugishita, T.; Murakami, T.; Murai, H.; Saikusa, T.; Horino, T.; Onoda, A.; Kajimoto, O.; Takahashi, R.; Takahashi, T. Effect of the defatted rice germ enriched with GABA for sleeplessness, depression, autonomic disorder by oral administration. *J. Jpn. Soc. Food Sci.* **2000**, *47*, 596–603. [CrossRef]
17. Ueno, H. Enzymatic and structural aspects on glutamate decarboxylase. *J. Mol. Catal. B-Enzym.* **2000**, *10*, 67–79. [CrossRef]
18. Coda, R.; Rizzello, C.G.; Gobbetti, M. Use of sourdough fermentation and pseudo-cereals and leguminous flours for the making of a functional bread enriched of γ-aminobutyric acid (GABA). *Int. J. Food Microbiol.* **2010**, *137*, 236–245. [CrossRef] [PubMed]
19. Rizzello, C.G.; Cassone, A.; Di Cagno, R.; Gobbetti, M. Synthesis of angiotensin I-converting enzyme (ACE)-inhibitory peptides and γ-aminobutyric acid (GABA) during sourdough fermentation by selected lactic acid bacteria. *J. Agric. Food Chem.* **2008**, *56*, 6936–6943. [CrossRef]
20. Villegas, J.M.; Brown, L.; de Giori, G.S.; Hebert, E.M. Optimization of batch culture conditions for GABA production by *Lactobacillus brevis* CRL 1942, isolated from quinoa sourdough. *LWT-Food Sci. Technol.* **2016**, *67*, 22–26. [CrossRef]
21. Lamberts, L.; Joye, I.J.; Beliën, T.; Delcour, J.A. Dynamics of γ-aminobutyric acid in wheat flour bread making. *Food Chem.* **2012**, *130*, 896–901. [CrossRef]
22. Coda, R.; Di Cagno, R.; Gobbetti, M.; Rizzello, C.G. Sourdough lactic acid bacteria: Exploration of non-wheat cereal-based fermentation. *Food Microbiol.* **2014**, *37*, 51–58. [CrossRef] [PubMed]

23. Schoenlechner, R.; Siebenhandl, S.; Berghofer, E. Pseudocereals. In *Gluten-Free Cereal Products and Beverages*, 1st ed.; Arendt, E.K., Dal Bello, F., Eds.; Academic Press: New York, NY, USA, 2008; p. 149-VI.
24. Chlopicka, J.; Pasko, P.; Gorinstein, S.; Jedryas, A.; Zagrodzki, P. Total phenolic and total flavonoid content, antioxidant activity and sensory evaluation of pseudocereal breads. *LWT-Food Sci. Technol.* **2012**, *46*, 548–555. [CrossRef]
25. Gorinstein, S.; Vargas, O.J.M.; Jaramillo, N.O.; Salas, I.A.; Ayala, A.L.M.; Arancibia-Avila, P.; Toledo, F.; Katrich, E.; Trakhtenberg, S. The total polyphenols and the antioxidant potentials of some selected cereals and pseudocereals. *Eur. Food Res. Technol.* **2007**, *225*, 321–328. [CrossRef]
26. Kalinova, J.; Dadakova, E. Rutin and total quercetin content in amaranth (*Amaranthus* spp.). *Plant. Food. Hum. Nutr.* **2009**, *64*, 68. [CrossRef]
27. Li, S.Q.; Zhang, Q.H. Advances in the development of functional foods from buckwheat. *Crit. Rev. Food Sci.* **2001**, *41*, 451–464. [CrossRef]
28. Lin, L.Y.; Liu, H.M.; Yu, Y.W.; Lin, S.D.; Mau, J.L. Quality and antioxidant property of buckwheat enhanced wheat bread. *Food Chem.* **2009**, *112*, 987–991. [CrossRef]
29. Paśko, P.; Sajewicz, M.; Gorinstein, S.; Zachwieja, Z. Analysis of selected phenolic acids and flavonoids in *Amaranthus cruentus* and *Chenopodium quinoa* seeds and sprouts by HPLC. *Acta Chromatogr.* **2008**, *20*, 661–672. [CrossRef]
30. Richardson, D.P. Wholegrain health claims in Europe. *Proc. Nutr. Soc.* **2003**, *62*, 161–169. [CrossRef]
31. Nimbalkar, M.S.; Pai, S.R.; Pawar, N.V.; Oulkar, D.; Dixit, G.B. Free amino acid profiling in grain Amaranth using LC–MS/MS. *Food Chem.* **2012**, *134*, 2565–2569. [CrossRef]
32. Jekle, M.; Houben, A.; Mitzscherling, M.; Becker, T. Effects of selected lactic acid bacteria on the characteristics of amaranth sourdough. *J. Sci. Food Agric.* **2010**, *90*, 2326–2332. [CrossRef]
33. Galli, V.; Venturi, M.; Pini, N.; Guerrini, S.; Granchi, L.; Vincenzini, M. Liquid and firm sourdough fermentation: Microbial robustness and interactions during consecutive backsloppings. *LWT-Food Sci. Technol.* **2019**, *105*, 9–15. [CrossRef]
34. Venturi, M.; Guerrini, S.; Granchi, L.; Vincenzini, M. Typing of *Lactobacillus sanfranciscensis* isolates from traditional sourdoughs by combining conventional and multiplex RAPD–PCR profiles. *Int. J. Food Microbiol.* **2012**, *156*, 122–126. [CrossRef] [PubMed]
35. Seseña, S.; Sánchez, I.; Palop, L. Characterization of *Lactobacillus* strains and monitoring by RAPD-PCR in controlled fermentations of "*Almagro*" eggplants. *Int. J. Food Microbiol.* **2005**, *104*, 325–335. [CrossRef]
36. Reguant, C.; Bordons, A. Typification of *Oenococcus oeni* strains by multiplex RAPD-PCR and study of population dynamics during malolactic fermentation. *J. Appl. Microbiol.* **2003**, *95*, 344–353. [CrossRef] [PubMed]
37. Alvarez-Jubete, L.; Wijngaard, H.; Arendt, E.K.; Gallagher, E. Polyphenol composition and in vitro antioxidant activity of amaranth, quinoa buckwheat and wheat as affected by sprouting and baking. *Food Chem.* **2010**, *119*, 770–778. [CrossRef]
38. Tuberoso, C.I.G.; Congiu, F.; Serreli, G.; Mameli, S. Determination of dansylated amino acids and biogenic amines in *Cannonau* and *Vermentino* wines by HPLC-FLD. *Food Chem.* **2015**, *175*, 29–35. [CrossRef] [PubMed]
39. Corona, O.; Alfonzo, A.; Ventimiglia, G.; Nasca, A.; Francesca, N.; Martorana, A.; Moschetti, G.; Settanni, L. Industrial application of selected lactic acid bacteria isolated from local semolinas for typical sourdough bread production. *Food Microbial.* **2016**, *59*, 43–56. [CrossRef]
40. Siragusa, S.; De Angelis, M.; Di Cagno, R.; Rizzello, C.G.; Coda, R.; Gobbetti, M. Synthesis of γ-aminobutyric acid by lactic acid bacteria isolated from a variety of Italian cheeses. *Appl. Environ. Microbiol.* **2007**, *73*, 7283–7290. [CrossRef]
41. Ayo, J.A. The effect of amaranth grain flour on the quality of bread. *Int. J. Food Prop.* **2001**, *4*, 341–351. [CrossRef]
42. Sanz-Penella, J.M.; Wronkowska, M.; Soral-Smietana, M.; Haros, M. Effect of whole amaranth flour on bread properties and nutritive value. *LWT-Food Sci. Technol.* **2013**, *50*, 679–685. [CrossRef]
43. Sindhuja, A.; Sudha, M.L.; Rahim, A. Effect of incorporation of amaranth flour on the quality of cookies. *Eur. Food Res. Technol.* **2005**, *221*, 597. [CrossRef]
44. Minervini, F.; Lattanzi, A.; De Angelis, M.; Di Cagno, R.; Gobbetti, M. Influence of artisan bakery-or laboratory-propagated sourdoughs on the diversity of lactic acid bacterium and yeast microbiotas. *Appl. Environ. Microbiol.* **2012**, *78*, 5328–5340. [CrossRef] [PubMed]

45. Ventimiglia, G.; Alfonzo, A.; Galluzzo, P.; Corona, O.; Francesca, N.; Caracappa, S.; Moschetti, G.; Settanni, L. Codominance of *Lactobacillus plantarum* and obligate heterofermentative lactic acid bacteria during sourdough fermentation. *Food Microbiol.* **2015**, *51*, 57–68. [CrossRef] [PubMed]
46. Lattanzi, A.; Minervini, F.; Di Cagno, R.; Diviccaro, A.; Antonielli, L.; Cardinali, G.; Cappelle, S.; De Angelis, M.; Gobbetti, M. The lactic acid bacteria and yeast microbiota of eighteen sourdoughs used for the manufacture of traditional Italian sweet leavened baked goods. *Int. J. Food Microbiol.* **2013**, *163*, 71–79. [CrossRef] [PubMed]
47. Vollmannova, A.; Margitanova, E.; Tóth, T.; Timoracka, M.; Urminska, D.; Bojňanská, T.; Čičová, I. Cultivar influence on total polyphenol and rutin contents and total antioxidant capacity in buckwheat, amaranth, and quinoa seeds. *Czech J. Food Sci.* **2013**, *31*, 589–595. [CrossRef]
48. Liukkonen, K.H.; Katina, K.; Wilhelmsson, A.; Myllymaki, O.; Lampi, A.M.; Kariluoto, S.; Piironen, V.; Heinonen, S.M.; Nurmi, T.; Adlercreutz, H.; et al. Process-induced changes on bioactive compounds in whole grain rye. *Proc. Nutr. Soc.* **2003**, *62*, 117–122. [CrossRef] [PubMed]
49. Poutanen, K.; Flander, L.; Katina, K. Sourdough and cereal fermentation in a nutritional perspective. *Food Microbiol.* **2009**, *26*, 693–699. [CrossRef]
50. Park, K.B.; Oh, S.H. Production of yogurt with enhanced levels of gamma-aminobutyric acid and valuable nutrients using lactic acid bacteria and germinated soybean extract. *Bioresour. Technol.* **2007**, *98*, 1675–1679. [CrossRef]
51. Kim, J.Y.; Lee, M.Y.; Ji, G.E.; Lee, Y.S.; Hwang, K.T. Production of γ-aminobutyric acid in black raspberry juice during fermentation by *Lactobacillus brevis* GABA100. *Int. J. Food Microbiol.* **2009**, *130*, 12–16. [CrossRef]
52. Thwe, S.M.; Kobayashi, T.; Luan, T.; Shirai, T.; Onodera, M.; Hamada-Sato, N.; Imada, C. Isolation, characterization, and utilization of γ-aminobutyric acid (GABA)-producing lactic acid bacteria from Myanmar fishery products fermented with boiled rice. *Fish. Sci.* **2011**, *77*, 279–288. [CrossRef]

foods

MDPI

Article

Total, Neutral, and Polar Lipids of Brewing Ingredients, By-Products and Beer: Evaluation of Antithrombotic Activities

Ronan Lordan *, Eoin O'Keeffe, Alexandros Tsoupras and Ioannis Zabetakis

Department of Biological Sciences, University of Limerick, Limerick V94 T9PX, Ireland; eoin.okeeffe@ul.ie (E.O.); Alexandros.Tsoupras@ul.ie (A.T.); Ioannis.Zabetakis@ul.ie (I.Z.)

* Correspondence: Ronan.Lordan@ul.ie; Tel.: +353-(0)61-202161

Received: 24 April 2019; Accepted: 19 May 2019; Published: 20 May 2019

Abstract: The in vitro antithrombotic properties of polar lipid constituents of malted grain (MG), pelleted hops (PH), brewer's spent grain (BSG), spent hops (SH), wort, and bottled beer from the same production line were assessed in human platelets. The total lipids (TL) were extracted according to the Bligh and Dyer method and further separated into the total neutral lipids (TNL) and total polar lipids (TPL) extracts by counter-current distribution. The TL, TNL, and TPL extracts of all samples were assessed for their ability to inhibit platelet-activating factor (PAF) and thrombin-induced human platelet aggregation. The raw materials, by-products, wort, and beer lipid extracts all exhibited antithrombotic properties against PAF and thrombin. However, the beer TPL exhibited the lowest IC_{50} values against PAF-induced (7.8 ± 3.9 μg) and thrombin-induced (4.3 ± 3.0 μg) platelet aggregation indicating that these polar lipids were the most antithrombotic. The lipid extracts tended to be more bioactive against the thrombin pathway. The fatty acid content of all the TPL extracts were assessed using GC-MS. The fatty acid composition of the most bioactive TPL extracts, the wort and the beer, shared similar fatty acid profiles. Indeed, it was noted that fermentation seems to play a role in increasing the antithrombotic properties of polar lipids against PAF and thrombin by moderately altering the polar lipid fatty acid composition. Furthermore, the use of brewing by-products as a source of functional cardioprotective lipids warrants further investigation and valorisation.

Keywords: fermentation; beer; brewer's spent grain; hops; polar lipids; platelet-activating factor; thrombin; cardiovascular disease; antithrombotic

1. Introduction

Cardiovascular diseases (CVD) are the leading cause of mortality globally, where diet and lifestyle are key modifiable risk factors [1]. The harmful effects of alcohol consumption have been well established, as excessive alcohol consumption has been linked to several chronic diseases, including cancer [2,3]. However, moderate alcohol consumption (up to 16 g alcohol/day) has been associated with cardiovascular health benefits, including reduced fasting glucose and blood insulin sensitivity versus non-alcoholic beer in healthy men [4]. Indeed, consumption of 1–2 alcoholic beverages/day is associated with reduced fibrinogen levels, reduced platelet aggregation, and an increase in high-density lipoprotein (HDL). These effects have mostly been attributed to the ethanol content and the presence of phenolic compounds [5–7]. Epidemiological studies demonstrate that moderate alcohol consumption reduced cardiovascular risk factors, morbidity, and mortality following a dose-effect relationship that is characterised by a J-shaped curve [6].

Atherosclerosis is the first step in the development of CVD [1]. Platelet-activating factor (PAF) is a potent platelet agonist and inflammatory mediator implicated in the onset and progression of atherosclerosis [8,9]. PAF and PAF-like molecules carry out their functions by binding to the

PAF-receptor (PAF-R), which is expressed in various cell types, including platelets, endothelial cells, neutrophils, and macrophages [1]. Activation of the PAF-R leads to the induction of multiple inflammatory pathways and platelet activation [1].

Thrombin is another important mediator of platelet activation [10]. Thrombin is a serine protease that participates in the coagulation cascade, activating factors V, VIII, XI, and XII, converting fibrinogen to fibrin, and activating other cell types [10,11]. Both PAF and thrombin are produced during coagulation and inflammation and play a crucial role in platelet activation and thrombus formation via G protein-coupled receptors [12]. Previously these two pathways were considered independent. However, recent evidence suggests that there is a crosstalk between coagulatory and inflammatory pathways during pathological processes, whereby inflammation leads to platelet activation in a reciprocal fashion [13]. Considering the important roles that PAF and thrombin hold at the nexus of coagulation and inflammation, antiplatelet and anti-inflammatory therapeutic and preventative strategies are required to prevent the development of chronic diseases such as CVD. Within this concept, several compounds of natural origin [1] can inhibit the binding of PAF to the PAF-R, which ameliorates the PAF inflammatory and prothrombotic response. Notably, polar lipid constituents of ale, lager, and stout have exhibited potent anti-PAF and antithrombotic properties [14]. However, little is known about the origin of these beer-derived antithrombotic polar lipid microconstituents and their effects against the thrombin pathway.

Beer is the most consumed alcoholic beverage globally. The raw materials used in beer production influence the different characteristics and properties of the beer [4]. Beers are produced from malted barley, water, hops (*Humulus lupulus*), and yeast (*Saccharomyces cerevisiae*). Barley (*Hordeum vulgare*) contains 2–4% (dry weight) lipid depending on various factors [15]. Commercial malts can contain up to 3.4% lipid. Approximately 70–90% of the fatty acid content of the barley and malt is triglycerides, 10–20% are sterol compounds, and approximately 10% are free fatty acids. The free lipid composition of the barley grain is approximately 68–75% neutral lipids, 7–26% glycolipids, and 9–18% phospholipids depending on the cultivar [16]. Germination of the barley and the mashing process can lead to the loss of lipid due to the release of fatty acids via the hydrolysis of triglycerides, which are then metabolised. The resulting fatty acids, mono-, and diglycerides do not tend to accumulate in the malt, and thus are not found in significant quantities in the finished product [17]. Moreover, several phospholipids can form complexes with amylose in starch before the brewing process [18]. A significant proportion of the lipid content is lost to the spent grains generated during the brewing process, and so the wort and beer contains low levels of lipid [19]. There is only a trace amount of lipid that remains in the final beer product, which are generally considered undesirable due to their impact on the formation of haze, the stability of beer foam, and the development of unfavourable flavours during conditioning [17,20].

Brewing yeasts also have the capacity to synthesise and alter several lipid species, including phospholipids and polyunsaturated fatty acids (PUFA) [21], which play a crucial role in the metabolic pathways and regulation of lipid catabolism and anabolism in yeast [17]. Furthermore, there is an increased concentration of stearic, *cis*-oleic, and linolenic acid in the wort as a result of endogenous lipase activity that releases free fatty acids from triglycerides and phospholipids in the mashing and malting process [21,22]. In addition, the composition of the malt and lauter turbidity can lead to the formation of triglycerides, diglycerides, monoglycerides, free fatty acids, phospholipids, and sphingolipids during the fermentation process [23].

The unfortunate consequence of beer production is the generation of brewing wastes and by-products, which are a significant environmental challenge, but may be valorised for the development of novel products [24]. Industrial-scale food processing by-products are increasingly being viewed as potential sources of bioactive ingredients. One such example in the brewing industry by-product previously sent to landfill is BSG [25], which is now primarily used for animal feed. The lipid content of BSG is approximately 10.0–13.5% (*w*/*w* of BSG samples) [26], where it is estimated that 9.1% of the total lipids are phospholipids [27].

Considering little is known about the fatty acid composition of bioactive polar lipid microconstituents of beer or brewing by-products, the aim of this study was to assess the antithrombotic properties and fatty acid composition of lipid extracts sampled at various stages of the brewing process, from the brewing raw materials (malted grain and hops), the by-products (BSG and spent hops), the wort, and the finished beer product from a commercial beer produced in Ireland.

2. Materials and Methods

2.1. Chemicals and Reagents

All glassware, chemicals, and solvents were of analytical grade and purchased from Fisher Scientific Ireland Ltd. (Dublin, Ireland). All reagents used for platelet aggregation, including bovine serum albumin (BSA), standard PAF, and standard thrombin, were high purity and purchased from Sigma-Aldrich (Wicklow, Ireland). Blood consumables, including needles (20G) and 8.2 mL sodium citrate S-monovettes, were purchased from Sarstedt Ltd. (Wexford, Ireland). For GC-MS, a pre-derivatised Supelco™ 37-component FAME standard mix, pre-derivatised heptadecanoic acid, and pre-derivatised heneicosanoic acid was purchased from Sigma-Aldrich (Wicklow, Ireland). All platelet aggregometry consumables were purchased from Labmedics LLP (Abingdon on Thames, UK). All GC-MS consumables were purchased from Apex Scientific Ltd. (Kildare, Ireland).

2.2. Beer Production and Samples Assessed

The beer, wort, raw materials, and brewing by-products used for this study were obtained from the Munster Brewery facility (Youghal, Co. Cork, Ireland). The samples assessed in this study were pelleted hops (PH), spent hops (SH), malted grains (MG), brewer's spent grains (BSG), wort, and the beer itself. All the raw materials, wort, and by-products obtained for this study correspond to the same batch and production line of the beer that was produced for and tested in this study. The beer is an organically produced Irish red ale for commercial sale under the name '12 Towers' brewed in accordance with organic standards certified by the Irish Organic Association.

The beer production is described in brief as follows. Of the malts used, approximately 90% of the overall malt was organic pale ale malt (Maris Otter malt) was kiln dried at 90–95 °C to produce the base malt of the beer with a maximum moisture content of 4.5% and a European Brewing Convention (EBC) colour scale between 7 and 10 (Castle Malting Ltd., Beloeil, Belgium). A small amount of organic roasted barley (approximately 10% of the total malt used), which was kiln dried to 230 °C that produced a roasted grain that had a maximum moisture content of 4.5% and was a colour rating between 1000–1400 EBC. Once milled by the brewery, the grains were mashed in the mash tun. To formulate the wort, 4 L of water was added per kg of crushed grain, which was steeped in the mash tun at 66 °C for 60 min. This process activated the α- and β-amylase enzymes (among others) that convert the starch to simple sugars to produce the so-called sweet wort. At the end of the mash, the lautering process begins by raising the temperature to 77–79 °C to denature the enzymes by adding heated sparge water, which also weakens the gravity of the wort runoff. This process takes approximately 3–4 h. Once Lautering was complete a sample of the BSG was stored. A brewer's hydrometer was used to measure the original gravity (OG) of the first running, which can be as high as 1.080 and also subsequently to confirm that the kettle wort is at an acceptable gravity prior to commencing boiling. The OG prior to fermentation was 1.042 and final gravity (FG) after fermentation was 1.010. These hydrometer readings are used to measure the fermentable and unfermentable substances the in wort before and during fermentation and to calculate the beers final alcoholic content (alcohol by volume or ABV) when OG is compared to the FG. Runnings from the mash tun were transferred to the kettle and once complete, the wort was heated to 100 °C and brought to the boil for 1 h, which sterilised the wort, denatured any remaining enzymes, and allowed for caramelisation of the sugars for flavour enhancement.

At the start of the boil a specific amount of organic pelleted hops (PH: Organic Goldings Hops, Charles Faram Ltd., Worcester, UK) was calculated based on flavour rating (approximately 200 g/hL) of the hops for each batch and was added to the kettle, which was boiled for a further hour. These pelleted hops are strobiles from the female hop plant that are air-dried and pelleted, and thus extremely concentrated in comparison to fresh strobiles. Thereafter, the wort underwent whirlpooling a rapid cooling process by being passed through a heat exchanger, which reduced the temperature of the wort to 20 °C. A sample of the wort was taken at this point. After whirlpooling when the wort had been moved on in the process, a sample of the SH was taken. The SH sample in reality is not purely hops, it is also known as the trub, which consists of primarily hop debris (10–20%), but is also a source of sedimented protein (50–70%), phenolic compounds (5–10%), carbohydrates (4–8%), and fatty acids (1–2%) originating from the hops and residues from the barley processing [28]. The cooled wort was transferred to the primary fermenter and was held at 20 °C. At this temperature, the yeast, a dehydrated *Saccharomyces cerevisiae* (Nottingham High Performance Ale Yeast, Lallemand Inc., Burton upon Trent, UK) was diluted in sterilised water and pitched (80–100 g per hL) into the filled fermentation tank. This is a top fermenting yeast that is commonly used to produce a wide variety of beers, including ambers, porters, stouts, and pale ales. The fermentation ran for four days. Following fermentation, a conditioning period was allowed to take place at 12 °C for approximately 7 days. After conditioning, the beer underwent cold crashing. In this process, the temperature of the conditioned beer is reduced to between 0–2 °C over 3–4 days. The process of cold crashing promotes the flocculation of yeast, which sink to the bottom of the tank due to gravity and are removed, thus increasing the clarity of the beer naturally without the need for additives. Indeed, cold crashing is also desirable due to its effects on other suspended particles, such as tannins, polyphenols, and proteins that can also flocculate and settle at the bottom of the tanks, preventing the undesirable phenomenon of chill haze. Once cold crashing was complete, the beer was bottled with a 2 g of 100% fermentable organic dextrose (Charles Faram Ltd., Worcester, UK) added to each bottle to encourage carbonation by any remaining yeast in the beer. The carbonation process takes approximately 14 days at room temperature until an optimal level of 2.2–2.5 volumes of CO_2 was achieved. Once complete the bottles were placed in cold storage (0–4 °C) for a short period of time until ready to distribute.

The finished bottled beer product, the wort, and the various by-products collected were placed in airtight containers and transported to the laboratory. The beer and wort were extracted on the day of arrival to the laboratory and the other by-products and raw materials were stored at −20 °C for a maximum of three weeks until required for extraction and analysis. All samples were taken in triplicate from different batches of the beer production process. The brewing process and sampling points are outlined in Figure 1.

2.3. Extraction and Isolation of the TL, TNL, and TPL Extracts

The total lipids (TL) of all samples were extracted in triplicate according to the Bligh and Dyer [29] method. Notably, the lipids from the brewing materials, by-products, wort, and beer were extracted from different batches of production. A tenth of each TL was stored under nitrogen at −20 °C and the remaining lipid was subjected to counter-current distribution as previously described [14] to obtain the total polar lipid (TPL) and the total neutral lipid (TNL) extracts. All extracts were stored under nitrogen at −20 °C until required for a maximum of 6 months as previously described [14].

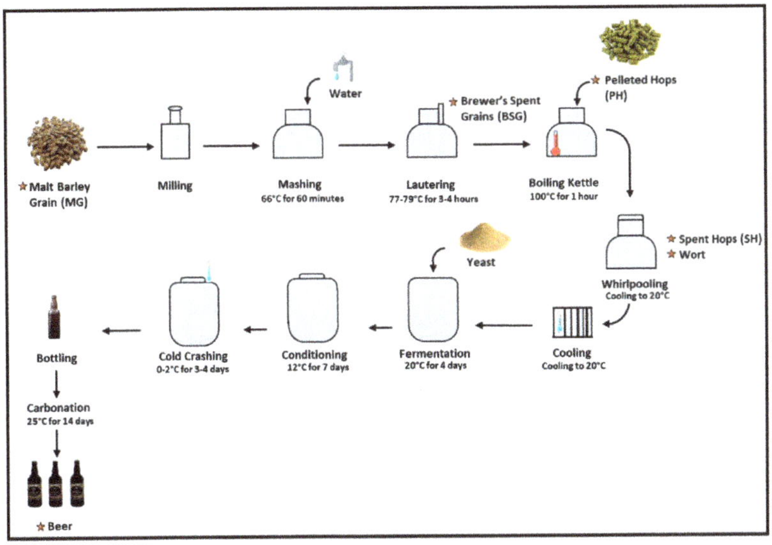

Figure 1. A schematic of the brewing process for the production of Irish red ale outlining the sampling points for the brewing materials, by-products, wort, and beer as highlighted with a star.

2.4. Platelet Aggregation Assay

The in vitro assessment of PAF and thrombin-induced platelet aggregation was carried out as previously described [14,30]. In brief, healthy human volunteers (N = 12) free from any form of antiplatelet therapy gave informed written consent and all protocols were executed in accordance with the Declaration of Helsinki following ethical approval by the University of Limerick Ethics Committee. Participants provided 50 mL or blood following an overnight fast (>8 h). The blood was drawn via venepuncture of the median cubital vein using a 20G needle into evacuated sodium citrate S-monovettes via the aspiration method (0.106 mol/L in a 1:10 ratio of citrate to blood). To obtain the platelet-rich plasma, S-monovettes were immediately centrifuged at 180× g for 18 min at 24 °C (Eppendorf 5702 R, Eppendorf Ltd, Stevenage, UK). A second centrifugation at 1500× g for 20 min at 24 °C was carried out to obtain the platelet-poor plasma (PPP). The PRP was standardised to 500,000 platelets μL^{-1} using a Shimadzu UV-1800 spectrophotometer (Kyoto, Japan), prior to analysis on a Chronolog-490 two channel platelet aggregometer (Chronolog, Havertown, PA, USA), coupled to the specialised AGGRO/LINK software package. Prior to testing, lipid samples and standard PAF were dissolved in a solution of BSA-saline (2.5 mg BSA/mL saline), while aliquots of stock solutions of active thrombin were diluted in appropriate amounts of physiological saline to obtain solutions of active thrombin with a final concentration in the aggregometer cuvette, ranging from 0.01 mU/mL NIH (National Institute of Health). The final concentration of PAF in the cuvette ranges from 1–5 × 10^{-8} M. Then, 250 µL of PRP was added to an aggregometer cuvette at 37 °C with stirring at 1000 rpm. The PRP was calibrated using the PPP as a blank. PAF and thrombin were added to the cuvettes in order to induce maximum reversible aggregation in the absence of any lipid samples. For each lipid sample, the mass of lipid required to inhibit 50% the PAF or thrombin-induced aggregation was calculated. Subsequently, the IC_{50} was calculated as previously described [14,30]. Platelet aggregation experiments for each TPL, TNL, and TL extract was carried out in triplicate against both PAF and thrombin as previously described [30,31].

2.5. Gas Chromatography-Mass Spectrometry

The preparation and analysis of the fatty acid methyl esters (FAME) were carried out in triplicate using 35 mg of the TPL samples of all samples as previously described [14]. Briefly, FAMEs were derivatised using a 0.5 M KOH CH_3OH 90% solution and extracted with n-hexane. The analysis was carried out using the internal standard method (Heneicosanoic acid—21:0) as previously described for other beverage analysis [14]. A five-point calibration curve was prepared using heneicosanoic acid (21:0 500 ppm injections) and five solutions of heptadecanoic acid (17:0—50, 100, 200, 400, and 800 ppm) methyl ester standards. Five 1 µL injections of each solution were analysed using a Varian 410-GC coupled to a Varian 210-MS equipped with a split/splitless injector (Agilent Technologies, Santa Clara, CA, USA). Separation of the FAME was conducted on an Agilent J&W DB-23 fused silica capillary column (60 m, '0.25 mm i.d.' 0.25 µm f.t.; Agilent Technologies). The ratio of the mean 17:0 to that of the internal standard (21:0) was used as the y-axis variable, while the concentration (ppm) of 17:0 was used as the x-axis variable of the calibration curve. The equation describing the curve was: $y = 0.0041x + 0.12$, with a $R^2 = 0.9969$, where the ratio of the area of the analyte peak to that of the internal standard represents the y value for the equation of the calibration curve and the x value represents the analyte concentration of a selected fatty acid in the lipid sample. The injector temperature of the Varian 410-GC and 210-MS was set at 230 °C with a split ratio of 1:20. The carrier gas was high purity helium with a liner flow rate of 1 mL/min. The oven temperature was initially programmed to 100 °C for 5 min, raised to 240 °C at 3 °C/min, and finally held isothermal at 240 °C for 10 min. Identification of FAME was achieved using a pre-derivatised standard 37-component FAME sample mix and comparison of the retention times and mass spectra of relative peaks with the aid of the Varian Star Chromatography Workstation Version 6 software (Agilent Technologies) and a NIST library of mass spectra (Gaithersburg, MD, USA). The percentage of each fatty acid was calculated using the peak area of the samples corrected by the respective response factors. Analyses were carried out in triplicate.

2.6. Statistical Analysis

All biological experimental analyses, extractions, and GC-MS analyses for each lipid sample were sample were completed in triplicate. The obtained results were expressed as a mean value ± standard deviation (SD). One-way analysis of variance (ANOVA) and the Tukey's honest significant difference (HSD) multiple comparison post-hoc test was used to determine the significant statistical differences between the analyses (SPSS Inc., Chicago, IL, USA).

3. Results

3.1. Lipid Extraction and Fractionation of Beer, and Brewing Materials

The TL, TNL, and TPL content of the beer and brewing materials are shown in Table 1 expressed in either mg/100g or g/100 g of TL, TNL, and TPL. The TNL and TPL are also expressed as a percentage of the TL. The MG and BSG were both considerably low in TL, but the TPL accounted for a substantial amount of the MG, of which just over half seems to be lost to the Brewer's spent grain in the brewing process. The largest amount of TL and TPL present in g/100 g of all samples was the PH, which was significantly higher than all the other samples. Notably the SH had a considerably lower quantity of TL and TPL than the PH, indicating that these lipids may have been extracted and diluted into the wort. However, both the wort and the beer contained extremely low amounts of lipid, the majority of which were polar lipids, results that are in accordance with previously published research [14,32,33].

Table 1. The total lipid (TL) and the total polar lipids (TPL) content of beer and brewing by-product are expressed as g/100 g and the total neutral lipid content (TNL) is expressed as mg/100 g. The TPL and TNL are also expressed as a percentage of the TL (mean ± SD, n = 3).

Sample	TL (g/100 g)	TNL (mg/100 g)	TNL (% TL)	TPL (g/100 g)	TPL (% TL)
MG	0.70 ± 0.10 ^{a}	100 ± 30 ^{a}	13.6 ± 2.9 ^{c}	0.52 ± 0.05 ^{a}	74.5 ± 3.5 ^{b}
BSG	1.05 ± 0.19 ^{a}	550 ± 110 ^{b}	52.6 ± 2.5 ^{e}	0.41 ± 0.09 ^{a}	38.9 ± 2.3 ^{a}
PH	14.17 ± 2.18 ^{b}	1630 ± 310 ^{c}	11.5 ± 3.7 ^{bc}	11.60 ± 1.68 ^{b}	79.4 ± 8.6 ^{bc}
SH	0.75 ± 0.06 ^{a}	160 ± 10 ^{a}	21.4 ± 3.0 ^{d}	0.55 ± 0.07 ^{a}	72.6 ± 4.1 ^{b}
Wort	0.03 ± 0.00 ^{a}	2.0 ± 1.0 ^{a}	5.4 ± 1.6 ^{ab}	0.03 ± 0.00 ^{a}	84.5 ± 8.9 ^{bc}
Beer	0.02 ± 0.00 ^{a}	0.3 ± 0.1 ^{a}	1.7 ± 0.4 ^{a}	0.02 ± 0.01 ^{a}	91.3 ± 2.7 ^{c}

^{a,b,c,d,e} Different superscripts indicate significant differences among different lipid extracts within the same lipid class ($p < 0.05$) when means are compared using a Tukey's HSD multiple comparison test. Abbreviations: BSG = brewer's spent grain; MG = malted grain; PH = pelleted hops; SH = spent hops.

3.2. Gas Chromatography-Mass Spectrometry Analysis

While the general lipid composition of beer, wort, barley, and malts have been comprehensively characterised by Bravi and colleagues [32–34], to the best of our knowledge the fatty acid profile of the polar lipids specifically have not been previously reported. Therefore, the fatty acid profile of each TPL extract were assessed by GC-MS (Table 2). Beer consists of many volatile and non-volatile compounds that affect the beer flavour and bioactivities [14,32]. Many of the volatiles and fatty acids in beer are synthesised by yeasts during fermentation, whereas others are derived from the raw materials [32]. Research demonstrates that the dominant fatty acids in the TL of barley grains, wort, and beer are palmitic, stearic, *cis*-oleic, linoleic, and α-linolenic acids [35]. However, little is known about the fatty acid profile of the polar lipid fractions of the raw materials, wort, or the beer itself.

In this study, the MG and BSG had similar TPL fatty acid profiles. Notably, there were statistically significant differences between the MG and BSG in the percentage of palmitic and linoleic acids, where both were higher percentages in the BSG than the MG TPL. Furthermore eicosenoic acid and docosahexaenoic acids were present as a low percentage of the MG but were not detected in the BSG TPL. Additionally, there were significant differences in the TPL fatty acid composition between the MG, BSG, and the beer, namely the TPL fatty acids differed by a higher percentage of SFA and a lower percentage of monounsaturated fatty acids (MUFA) and polyunsaturated fatty acids (PUFA) in the beer TPL in contrast to the MG and BSG. While the MUFA and PUFA were present in lower amounts in the beer compared to the MG and BSG, arachidonic acid was present in the beer in notably high amounts (4.9%) but was not present in the MG or BSG. Furthermore, myristic and stearic acid were also present in a significantly high percentage of the beer TPL in contrast to the MG and BSG. The MG and the wort also differ in composition, as there was a higher percentage of palmitic, stearic, arachidonic, and eicosapentaenoic acids and a lower percentage of linoleic, arachidic, eicosenoic acid, and docosahexaenoic acid with no phytochemicals detected in the MG, BSG, or wort TPL.

Table 2. The fatty acid profile and volatile compounds detected in the total polar lipid (TPL) extracts of each sample are expressed as a percentage of the total volatile components detected by GC-MS (mean ± SD, n = 3).

Fatty Acids		Malt Grain	Spent Grain	Pelleted Hops	Spent Hops	Wort	Beer
8:0	Caprylic acid	ND	ND	0.04 ± 0.01^{b}	0.02 ± 0.01^{ab}	0.01 ± 0.00^{a}	ND
10:0	Capric acid	ND	ND	0.06 ± 0.00^{b}	ND	0.01 ± 0.00^{a}	0.02 ± 0.01^{a}
12:0	Lauric acid	0.17 ± 0.02^{d}	ND	0.02 ± 0.00^{a}	0.03 ± 0.01^{a}	0.09 ± 0.01^{c}	0.07 ± 0.00^{b}
12:1	*cis*-Lauroleic acid	ND	ND	0.05 ± 0.01	ND	ND	ND
14:0	Myristic acid	0.62 ± 0.06^{b}	0.69 ± 0.06^{b}	0.33 ± 0.05^{a}	1.04 ± 0.11^{c}	1.58 ± 0.04^{d}	1.55 ± 0.09^{d}
14:1	*cis*-Myristoleic acid	ND	0.41 ± 0.11	ND	ND	ND	ND
15:0	Pentadecylic acid	0.31 ± 0.04^{a}	0.22 ± 0.07^{a}	0.49 ± 0.14^{b}	0.25 ± 0.01^{a}	0.17 ± 0.02^{a}	0.16 ± 0.02^{a}
16:0	Palmitic acid	19.83 ± 0.93^{a}	27.86 ± 0.80^{b}	22.05 ± 1.23^{a}	30.38 ± 1.0^{c}	31.80 ± 0.60^{c}	32.34 ± 0.68^{c}
16:1	*cis*-Palmitoleic acid	0.39 ± 0.04^{a}	0.30 ± 0.12^{a}	2.35 ± 0.26^{c}	1.15 ± 0.15^{b}	0.60 ± 0.02^{a}	0.33 ± 0.21^{a}
17:0	Margaric acid	0.18 ± 0.02^{a}	ND	1.24 ± 0.03^{d}	0.46 ± 0.03^{c}	0.24 ± 0.02^{b}	0.23 ± 0.01^{ab}
17:1	*cis*-Heptadecenoic acid	0.09 ± 0.01^{a}	ND	0.75 ± 0.04^{d}	0.36 ± 0.03^{c}	0.20 ± 0.01^{b}	0.20 ± 0.03^{b}
18:0	Stearic acid	2.63 ± 0.62^{ab}	2.23 ± 0.25^{a}	2.85 ± 0.08^{abc}	3.85 ± 0.18^{d}	3.61 ± 0.09^{cd}	3.32 ± 0.10^{bcd}
18:1 c9	*cis*-Oleic acid	9.04 ± 0.19^{d}	8.81 ± 0.44^{d}	4.39 ± 0.11^{a}	6.67 ± 0.26^{d}	6.12 ± 0.24^{bc}	5.62 ± 0.18^{b}
18:1 t13	*trans*-Oleic acid	0.66 ± 0.02^{a}	1.01 ± 0.09^{ab}	1.27 ± 0.16^{bc}	1.82 ± 0.24^{d}	1.42 ± 0.09^{c}	1.26 ± 0.14^{bc}
18:2 c9, c12	Linoleic acid	56.67 ± 0.77^{e}	51.83 ± 1.59^{d}	25.46 ± 1.5^{a}	40.68 ± 0.34^{b}	44.78 ± 0.06^{c}	43.48 ± 1.55^{bc}
18:3 c6, c9, c12	γ-Linolenic acid	ND	ND	0.58 ± 0.02^{b}	0.12 ± 0.01^{a}	ND	ND
18:3 c9, c12, c15	α-Linolenic acid	6.80 ± 0.95^{abc}	5.87 ± 0.54^{ab}	23.42 ± 1.4^{d}	8.83 ± 0.92^{c}	7.72 ± 0.71^{bc}	5.13 ± 0.26^{a}
20:0	Arachidic acid	0.58 ± 0.10^{ab}	0.78 ± 0.17^{bc}	1.02 ± 0.04^{c}	0.48 ± 0.07^{a}	ND	ND
20:1 c13	Eicosenoic acid	ND	ND	0.25 ± 0.05^{a}	0.54 ± 0.04^{c}	0.39 ± 0.02^{b}	ND
20:2 c11, c14	Eicosadienoic acid	ND	ND	1.00 ± 0.05^{c}	0.49 ± 0.07^{b}	0.27 ± 0.01^{a}	0.31 ± 0.04^{a}
20:4 c5, c8, c11, c14	Arachidonic acid	ND	ND	ND	ND	0.52 ± 0.06	4.93 ± 0.02
20:5 c5, c8, c11, c14, c17	Eicosapentaenoic acid	ND	ND	0.53 ± 0.01^{b}	0.16 ± 0.03^{a}	0.41 ± 0.10^{b}	ND
22:0	Behenic acid	0.30 ± 0.04^{a}	ND	1.32 ± 0.32^{c}	0.72 ± 0.13^{b}	0.41 ± 0.02^{ab}	0.29 ± 0.07^{a}
22:1	Erucic acid	0.32 ± 0.09^{a}	ND	0.41 ± 0.06^{a}	0.34 ± 0.09^{a}	ND	ND
22:6 c4, c7, c10, c13, c16, c19	Docosahexaenoic acid	0.47 ± 0.09^{a}	ND	1.46 ± 0.17^{b}	0.43 ± 0.13^{a}	ND	ND
Σ_{SFA}		24.43 ± 0.70^{a}	31.79 ± 0.94^{b}	29.40 ± 1.32^{b}	37.23 ± 0.91^{c}	37.94 ± 0.60^{c}	37.97 ± 0.54^{c}
Σ_{MUFA}		10.67 ± 0.14^{c}	10.53 ± 0.41^{c}	9.47 ± 0.48^{b}	10.86 ± 0.37^{c}	9.13 ± 0.21^{b}	7.42 ± 0.45^{a}
Σ_{PUFA}		63.95 ± 1.53^{c}	57.69 ± 1.09^{b}	52.87 ± 2.66^{a}	50.71 ± 0.67^{a}	53.69 ± 0.60^{a}	53.75 ± 1.76^{a}

Table 2. *Cont.*

Fatty Acids	Malt Grain	Spent Grain	Pelleted Hops	Spent Hops	Wort	Beer
Volatiles						
Hexanedioic acid	ND	ND	ND	ND	0.28 ± 0.04	1.12 ± 0.21
Aromadendrene oxide	ND	ND	1.77 ± 0.39	0.11 ± 0.02	ND	ND
2,4-Di-*tert*-butylphenol	ND	ND	ND	ND	ND	0.12 ± 0.01
β-Caryophyllene	ND	ND	2.02 ± 0.37 [b]	0.37 ± 0.05 [a]	ND	0.07 ± 0.01 [a]
2-Dodecanone	ND	ND	0.07 ± 0.05	0.04 ± 0.01	ND	ND
Cubenol	ND	ND	0.24 ± 0.18	ND	ND	ND
Tau-Cadinol	ND	ND	0.14 ± 0.08	ND	ND	ND
Tau-Muurolol	ND	ND	0.29 ± 0.02	ND	ND	0.08 ± 0.00
$\Sigma_{Volatiles}$	ND	ND	8.90 ± 0.32 [b]	0.95 ± 0.32 [a]	0.28 ± 0.04	1.37 ± 0.22 [a]

[a,b,c,d,e] Mean values ± SD (n = 3), different letters in the same row indicate statistically significant differences between the lipid compositions when means are compared using Tukey's HSD multiple comparison test ($p \leq 0.05$). Abbreviations: c = *cis*; MUFA = monounsaturated fatty acids; ND: non-detectable; PUFA = polyunsaturated fatty acids; SFA = saturated fatty acids; t = *trans*.

Another contributing factor to the fatty acid composition of beer is the hops. Hops are known for their high concentrations of volatile components that can impart bitter flavours and aroma to the wort and final product in the brewing process [36]. However, there is relatively little literature addressing the fatty acid compositions of the TL or TPL of hops. In this study, there were significant differences between the fatty acid compositions of the TPL of the PH and the SH. While the total percentage of PUFA in the TPL of the PH and SH were similar, there was a significant increase in the percentage of SFA and MUFA in the SH. However, this is most likely not due to an actual rise in the levels of these fatty acids in the TPL, but may be due to the fact that the PH TPL extract contained a high amount of volatile components that seemed to significantly reduce in the SH. Considering these levels of volatile phytochemicals were not present in the wort or beer in any significant percentages, it is likely that these volatiles were lost during the boiling of the wort, which has been documented previously in studies [37]. Generally, there was a lower percentage of palmitoleic, margaric, and α-linolenic, arachidic, eicosadienoic, docosahexaenoic acids in the SH in contrast to the PH, whereas there was a significantly higher percentage of myristic, palmitic, stearic, *cis*-oleic, *trans*-oleic, and linoleic acids in the SH in contrast to the PH. Furthermore, there was a high percentage of caryophyllene and aromadendrene oxide in the PH TPL.

While the intention of this analysis was to assess the fatty acid profile of the TPL extract, volatile phytochemicals and other constituents were also detected in the TPL extracts of beer, PH, and SH in low percentages. The beer TPL extract contained the highest percentage of phytochemicals at 1.37%, which can mainly be attributed to the detection of adipic acid (or hexanedioic acid). Other compounds identified in the final beer TPL extract were β-caryophyllene, 2,4-Di-*tert*-butylphenol, and tau-muurolol in low percentages. β-caryophyllene and other essential oils were also present in the PH TPL extracts, including tau-muurolol, a cadinene sesquiterpenoid that is a plant metabolite, fungicide, and volatile oil [38], which is commonly found in hops [39]. The PH also contains β-caryophyllene, a bicyclic sesquiterpene that was also present in the beer and SH. β-caryophyllene is a characteristic essential oil of hops that is usually in high abundance in comparison to other phytochemicals [36,40] and is the most abundant phytochemical in the PH and SH TPL extract.

3.3. Platelet Aggregation Assay Analysis

The TL, TNL, and TPL of each extract were assessed for their ability to inhibit PAF and thrombin-induced platelet aggregation in human PRP. The results of the platelet aggregometry assay for all lipid samples are expressed as an IC_{50} value, which is the mass of the lipid sample in micrograms (μg) required to inhibit half (50%) the maximum-reversible PAF/thrombin-induced platelet aggregation. The IC_{50} results for each extract tested against PAF-induced platelet aggregation are presented in Table 3. It is clear from the data presented that overall the TL and TPL are the most bioactive fractions in the various brewing-related extracts. Overall, the TNL extracts generally exhibited poor bioactivity. Notably in the MG, the TNL was slightly more bioactive than the TL, but was less bioactive than the TPL. Furthermore, the BSG TL had an exceedingly low IC_{50}, and thus a higher anti-PAF effect, in comparison to the relative effect of the TNL and TPL. It is not known why this may be the case and further research is required to ascertain whether there may be synergistic effects between the compounds extracted in the TL of the BSG. In terms of the hop extracts, TL, TNL, and TPL IC_{50} values were similar and did not show any statistically significant differences between the PH and the SH. The wort and final beer product were generally the most bioactive TL, TNL, and TPL extracts, although not statistically significantly different from each other.

The IC_{50} results for each lipid extract assessed against thrombin-induced platelet aggregation are presented in Table 4. Similarly to the results of the PAF-induced platelet aggregation assay, it is clear that overall the TL and TPL are the most bioactive fractions in the various brewing-related extracts. The IC_{50} values of the TNL of the MG, BSG, PH, and SH were high, indicating poor inhibition against thrombin. However, the TNL IC_{50} values for the wort and beer were comparable to each other and were considerably lower, and thus more effective against thrombin than the TNL extracts from all other sources. Like the TNL extracts, the TPL of the MG, BSG, PH, and SH were high, indicating poor thrombin inhibition. However, the wort and the beer exhibited extremely low IC_{50} values that were not statistically significantly different from each other. Notably, the TL extracts of the MG, BSG, PH, and SH also possess considerable inhibitory properties against thrombin, although not as potent as the beer or the wort.

Table 3. The in vitro biological activities of the total lipids (TL), total neutral lipids (TNL), and total polar lipids (TPL) of the beer and brewing by-products against platelet-activating factor (PAF)-induced human platelet aggregation, expressed as an IC_{50} in micrograms (μg) of the sample extract. The hPRP concentration was approximately 500,000 platelets μL^{-1}. The final concentration of PAF in the cuvette was 2.6×10^{-8} M. All experimental analyses were carried out in triplicate (mean ± SD, $n = 3$).

Sample	TL	TNL	TPL
MG	495 ± 105^{b}	298 ± 89^{a}	191 ± 58^{ab}
BSG	69 ± 33^{a}	610 ± 136^{b}	617 ± 184^{c}
PH	453 ± 109^{b}	1088 ± 172^{c}	473 ± 280^{c}
SH	519 ± 81^{b}	924 ± 166^{c}	436 ± 142^{bc}
Wort	70 ± 29^{a}	175 ± 61^{a}	58 ± 11^{a}
Beer	6.4 ± 4.5^{a}	248 ± 66^{a}	7.8 ± 3.9^{a}

^{a,b,c} Different superscripts indicate significant differences among different lipid extracts within the same lipid class ($p < 0.05$), when means are compared using ANOVA and Tukey's HSD multiple comparison test. Abbreviations: BSG = brewer's spent grain; hPRP = human platelet-rich plasma; MG = malt grain PAF = platelet-activating factor; PH = pelleted hops; SH = spent hops; TL = total lipids; TNL = total neutral lipids; TPL = total polar lipids.

Table 4. The in vitro biological activities of the total lipids (TL), total neutral lipids (TNL), and total polar lipids (TPL) of the beer and brewing by-products against thrombin-induced human platelet aggregation. Results are expressed as an IC_{50} in micrograms (μg) of each lipid extract. The hPRP concentration was approximately 500,000 platelets μL^{-1}. The final concentration of thrombin in the cuvette was 0.1–1.0 mU/mL. All experimental analyses were carried out in triplicate (mean ± SD, $n = 3$).

Sample	TL	TNL	TPL
MG	112 ± 21^{b}	433 ± 77^{b}	247 ± 39^{b}
BSG	87 ± 10^{b}	409 ± 30^{b}	203 ± 49^{b}
PH	221 ± 42^{c}	478 ± 97^{b}	207 ± 51^{b}
SH	155 ± 56^{bc}	572 ± 76^{b}	396 ± 62^{c}
Wort	10 ± 3.7^{a}	165 ± 61^{a}	24 ± 17^{a}
Beer	2.4 ± 0.9^{a}	206 ± 73^{a}	4.3 ± 3.0^{a}

^{a,b,c} Different superscripts indicate significant differences among different lipid extracts within the same lipid class ($p < 0.05$), when means are compared using ANOVA with Tukey's HSD multiple comparison test. Abbreviations: BSG = brewer's spent grain; hPRP = human platelet-rich plasma; PAF = platelet-activating factor; MG = malt grain; PH = pelleted hops; SH = spent hops; TL = total lipids; TNL = total neutral lipids; TPL = total polar lipids.

4. Discussion

Previous research has demonstrated that commercial ale, lager, and stout possess potent anti-PAF activities as demonstrated through PAF-induced platelet aggregation assays [14]. Therefore, the aim of this study was to assess the antithrombotic activities and fatty acid composition of brewing raw materials, by-products, wort, and beer from a single production line in an active brewery.

In this study the TL, TNL, and TPL were extracted from the raw materials (MG and PH), by-products (BSG and SH), wort, and Irish red ale, all originating from the same production line. The raw materials used in beer production contain a significant amount of lipids, particularly from the malted barley and hops. However, only trace amounts remain in the final beer product [19]. The lipid content of the MG mix used as a raw material in this study was considerably low (0.7 g/100 g), where other ale malts can possess higher lipid levels between 2.8–3.4 g/100 g as reported by Anness [41]. There are several reasons why the lipid content of the MG was low in this study, including that the lipids of barley tend to form complex interactions making them tightly bound to starch [18], thus making them difficult to extract. It is possible that much of the lipid was not efficiently extracted using the Bligh and Dyer [29] method. This extraction method does not use harsh acid or heat treatments and therefore does not efficiently extract all of the lipids but was chosen for this study as an efficient method for extracting bioactive lipids against platelet aggregation. Indeed, studies have demonstrated a large variation of lipid yield between different extraction procedures, including some being more efficient than others for co-extracting non-lipid substances as part of the crude lipid content [42]. Other studies have also acknowledged that barley lipids are poorly extractable without the use of hot alcoholic extraction [43]. Therefore, further studies should consider the use of other extraction methods.

Notably, the BSG had a non-statistically significant higher lipid content than the malt, but a statistically lower TPL content (~39% of the total lipids). This is in accordance with previous research that estimates that 30% of the barley lipid content is lost during the germination of barley due to the hydrolysis of triglycerides, which are subsequently metabolised [41].

The greatest amount of lipid extracted from any sample was the PH (14.7 g/100 g) with 79% being polar compounds. In hops generally, there is a limited amount of fatty acids (1–2%) present [28]. However, the essential oil content of air-dried female hop flowers (strobiles) is generally around 0.5–3%, whereas waxes and steroids are generally present in trace to high amounts (25%) [44]. Considering that the hops used in this study were dried and pelleted, the oil and fatty acid content was highly concentrated, thus explaining the high lipid content as per Table 1. The hops were added during the boiling process to allow the essential oils present in the hops to contribute to beer flavour and aroma [40]. However, much of the essential oils present in these bittering hops are highly volatile and some are lost during the boiling process to evaporation [37], while any that remain are dispersed in the wort, but may be filtered out during clarification of the wort. At the end of the boiling process, the SH are collected within the trub. As demonstrated in Table 1, the SH contains low levels of lipids. This is most likely due to the pelleted hops being dispersed and rehydrated within the wort and the loss of some volatile compounds from the essential oils during the boiling process. Notably, there was a high percentage of caryophyllene and aromadendrene oxide in the PH TPL. Aromadendrene oxide is an oxygenated sesquiterpene that is considered an essential oil, which exhibits anticancer properties [45].

During the brewing process, a significant proportion of the lipid content is lost to the spent grains [17]. Indeed, in this study the lipid content of the beer and wort were considerably low, but the low beer TL is in accordance with previously published research [14,41]. The presence of lipids in beer is generally considered as a negative proponent due to their effect on foam stabilisation and flavour. Therefore, the reduction of lipid levels is actively reduced where possible through various parts of the clarification processes used by breweries. The low lipid content of beer generally may also be due to the fact that barley contains polar lipids and fatty acids that are closely associated with polysaccharides that create amylose-lipid complexes that are difficult to fully extract [23]. Considering the high TL and TPL content of the BSG, it is likely that this is the case and that these lipids are lost to the BSG.

The fatty acid composition of the TPL extracts of all the brewing ingredients, by-products, wort and beer was determined. While the intention of the GC-MS analysis was to assess the fatty acid profile of the TPL extract, volatile phytochemicals and other constituents were also detected in the TPL extracts of the beer, PH, and SH in low but considerable percentages. The beer TPL extract contained the highest percentage of phytochemicals at 1.37%, which can mainly be attributed to the detection of adipic acid, an unusual non-volatile, alcohol soluble, dicarboxylic acid found sparsely in nature but is used as a food additive (E355) as a firming or raising agent, which has tart flavour and is safe for human consumption in low doses [46]. As it was not an intentional additive in the production of the beer, it is not known where in the brewing process the adipic acid originates from or if it was a contaminant as it was only found in the TPL of the wort and the final beer product.

Phenolic compounds play a significant role in aroma and flavour development in beer production [47]. Present in beer, 2,4-Di-*tert*-butylphenol is a phenolic compound that is produced by a variety of plants, but can also synthesised enzymatically by *S. cerevisiae* from organic acids [48]. Interestingly it is a bioactive compound with potential anticancer effects [49], antioxidant activities, and may be preventative against the neuroinflammatory effects of amyloid beta (Aβ) in animal models of Alzheimer's disease [50]. Considering some phenolic compounds have antiplatelet properties [51], it is yet to be determined whether 2,4-Di-*tert*-butylphenol contributed to the antithrombotic activities observed in this study. Tau-muurolol was also detected, which is a cadinene sesquiterpenoid that is a plant metabolite, fungicide, and volatile oil [52] that was detected in the beer and PH TPL extracts. β-caryophyllene, a bicyclic sesquiterpene was also present in the beer, PH, and SH. β-caryophyllene is a characteristic essential oil of hops that is usually in high abundance in comparison to other phytochemicals [36,40] and is the most abundant phytochemical in the PH and SH TPL extract. Notably, caryophyllene compounds may possess anticancer, analgesic, antioxidant, antimicrobial, and anti-inflammatory activities [53]. Indeed, caryophyllene molecules were present in abundance in essential oil extracts from 25 species of plants that demonstrated anti-platelet activity against adenosine diphosphate (ADP), arachidonic acid, and the thromboxane A_2 agonist U46619-induced platelet aggregation in guinea pig and rat plasma [54].

PAF and thrombin-induced platelet aggregation assays were used to assess the antithrombotic activity of the TL, TNL, and TPL extracts of the brewing raw materials, by-products, wort, and beer. Generally, the TPL extracts were the most potent against PAF-induced platelet aggregation. The TL extracts exhibited considerable anti-PAF and anti-thrombin effects, but generally the TPL extracts were more potent against PAF, whereas in the case of thrombin, the TL in cases were considerably more antithrombotic than the TNL and moderately more potent that the TPL extracts. Considering, the overall poorer inhibitory effects of both the TNL and TPL extracts of the MG, BSG, PH, and SH against thrombin, the compounds present in the combined TL extract may induce synergistic effects that improve the antithrombotic properties of these extracts, as has previously been demonstrated in beer [14] and other extracts of natural origin marine extracts [55]. It can be suggested that coextracted microconstituents such as phenolic compounds and phytochemicals with potential antithrombotic activities may in part be responsible for these observations as previously demonstrated [51]. Indeed, considering the BSG possessed potent anti-PAF and anti-thrombin activities, these by-products of the brewing industry could potentially be used in the development of nutraceuticals or functional foods and animal feeds, as has previously been demonstrated using by-products of the olive oil industry [56]. Indeed, BSG contains other significant bioactive microconstituents such as peptides and phenolic compounds that exhibit antioxidant activity [26] that supports the need for further research into the valorisation of BSG as a functional product for human health.

The beer and wort extracts were the most bioactive fractions against PAF and thrombin. While not deemed statistically significantly different, the bioactivity of the TL and TPL of the wort seemed to increase considerably following fermentation. Previous studies in dairy products have shown that the fermentation process may play a role in the biosynthesis of functional antithrombotic lipids [57]. However, it has yet to be confirmed whether *S. cerevisiae* can indeed biosynthesise antithrombotic polar lipids, but this study does demonstrate that yeasts may affect the fatty acid composition of the polar lipids. Further structure activity relationship studies are required to confirm this notion.

As depicted in Table 2, there is a statistically significant increase in the percentage of polar lipids bearing fatty acids arachidonic acid and eicosapentaenoic acid (EPA) in their structures as a result or the wort fermentation. Various studies have demonstrated that polar lipids of natural origin that bear these fatty acids in their structures along with stearic, cis-oleic, and linoleic acids, all of which are present in abundance in the beer and wort TPL, exhibit potent antithrombotic properties against PAF-induced platelet aggregation [9]. Similarly, the beer (red ale) TL and TPL IC_{50} values obtained in this study were similar to those published for Smithwick's red ale TL and TPL against PAF-induced platelet aggregation using the same methods [27]. The fatty acid compositions of the wort and beer TPL share structural resemblance to the classical PAF structure, which is generally composed of palmitic (68%), stearic (27%), or oleic (4%) acids at the *sn*-1 position, with acetic acid esterified to the *sn*-2 position, and a phosphocholine group at the *sn*-3 position [58]. Juxtaposed, the most abundant fatty acids of the TPL of the wort were palmitic (31.8%), stearic (3.6%), oleic (6.1%), and linoleic (44.8%)

acids and the most abundant fatty acids present in the TPL of the beer were palmitic (32.3%), stearic (3.3%), oleic (5.6%), and linoleic (43.4%) acids. Further research is required to confirm whether there is structural homology between some polar lipids and PAF, which facilitates their binding to the PAF-R, which may account for their potent biological actions against PAF.

In contrast to the wealth of evidence demonstrating that polar lipids can inhibit the biological actions of PAF, there is little published research demonstrating the mechanisms for the antiplatelet effects of food-derived polar lipids against thrombin-induced platelet aggregation. It seems that the amphiphilic properties of these bioactive polar lipid moieties expedite their transfer from blood lipoproteins to the membranes of circulating platelets. Thus such bioactive polar lipids can either directly affect several platelet membrane receptors related to platelet activation (i.e. binding of polar lipids to the PAF-R) [31] or indirectly affect these platelet receptors. Polar lipids may indirectly affect platelet receptors through altering the microenvironment and polarisation of the phospholipid membrane, which potentially alters the affinity of a ligand to a receptor relating to platelet activation, such as thrombin [9,31,59,60]. However, further research is required to verify these potential mechanisms and to discern the structures of these compounds, in order to fully elucidate the structure activity relationships between bioactive polar lipid extracts and their overall antiplatelet effects.

5. Conclusions

The antithrombotic activities of lipid extracts from brewing raw materials, by-products, wort, and beer were assessed. The most bioactive anti-PAF and anti-thrombin polar lipid extracts originated from the wort and the final beer product. While not statistically significantly different, it is apparent that fermentation of the wort may play a key role in increasing the anti-PAF bioactivity of polar lipids extracted from beer. These findings are in accordance with previous studies demonstrating that fermentation plays a key role in altering the bioactivity of anti-PAF polar lipids during milk fermentation. Indeed, this research supports and furthers previously published research demonstrating the presence of potent anti-PAF polar lipids in red ale. It was also observed that some phytochemicals and phenolic compounds may contribute to antithrombotic properties of these lipid extracts. Furthermore, this is the first study to demonstrate the anti-thrombin activities of beer polar lipids, but further research is required to discern the exact structures and mechanisms responsible for these observations. Moreover, it was determined that the BSG may be a suitable brewing industry by-product for valorisation as potential nutraceuticals or functional foods for improved human cardiovascular health.

Author Contributions: The study was conceptualized by I.Z., R.L., and A.T. The methods were developed by R.L. and A.T.; All experiments and interpretation of results were performed by E.O., R.L., and A.T. The manuscript was prepared by R.L. and A.T. and all authors reviewed and approved the final submission.

Funding: This research received no external funding.

Acknowledgments: The authors would like to express their sincere gratitude to Padraig Hyde and Adrian Hyde of Munster Brewery for their kind donation of their time, expertise, beer, and brewing materials for this study. The authors would also like to thank all the volunteers for their donations, Elaine Ahern for her phlebotomy support, and the Department of Biological Science, University of Limerick, Ireland for their continued support.

Conflicts of Interest: The authors declare no conflicts of interest but admit to thoroughly enjoying the final product.

References

1. Tsoupras, A.; Lordan, R.; Zabetakis, I. Inflammation, not cholesterol, is a cause of chronic disease. *Nutrients* **2018**, *10*, 604. [CrossRef] [PubMed]
2. Schwarzinger, M.; Thiébaut, S.P.; Baillot, S.; Mallet, V.; Rehm, J. Alcohol use disorders and associated chronic disease—A national retrospective cohort study from france. *BMC Public Health* **2017**, *18*, 43. [CrossRef]
3. Lordan, R.; Tsoupras, A.; Zabetakis, I. The potential role of dietary platelet-activating factor inhibitors in cancer prevention and treatment. *Adv. Nutr.* **2019**, *10*, 148–164. [CrossRef] [PubMed]

4. Nogueira, L.C.; do Rio, R.F.; Lollo, P.C.B.; Ferreira, I.M.P.L.V.O. Moderate alcoholic beer consumption: The effects on the lipid profile and insulin sensitivity of adult men. *J. Food Sci.* **2017**, *82*, 1720–1725. [CrossRef] [PubMed]
5. Arranz, S.; Chiva-Blanch, G.; Valderas-Martínez, P.; Medina-Remón, A.; Lamuela-Raventós, R.M.; Estruch, R. Wine, beer, alcohol and polyphenols on cardiovascular disease and cancer. *Nutrients* **2012**, *4*, 759. [CrossRef]
6. De Gaetano, G.; Costanzo, S.; Di Castelnuovo, A.; Badimon, L.; Bejko, D.; Alkerwi, A.; Chiva-Blanch, G.; Estruch, R.; La Vecchia, C.; Panico, S.; et al. Effects of moderate beer consumption on health and disease: A consensus document. *Nutr. Metab. Cardiovasc. Dis.* **2016**, *26*, 443–467. [CrossRef]
7. Lombardi, G.; Vannini, S.; Blasi, F.; Marcotullio, M.C.; Dominici, L.; Villarini, M.; Cossignani, L.; Moretti, M. In vitro safety/protection assessment of resveratrol and pterostilbene in a human hepatoma cell line (HepG2). *Nat. Prod. Commun.* **2015**, *10*, 1403–1408. [CrossRef]
8. Lordan, R.; Nasopoulou, C.; Tsoupras, A.; Zabetakis, I. The anti-inflammatory properties of food polar lipids. In *Bioactive Molecules in Food*; Mérillon, J.M., Ramawat, K.G., Eds.; Springer International Publishing: Cham, Switzerland, 2018; pp. 1–34.
9. Lordan, R.; Tsoupras, A.; Zabetakis, I. Phospholipids of animal and marine origin: Structure, function, and anti-inflammatory properties. *Molecules* **2017**, *22*, 1964. [CrossRef]
10. Jaberi, N.; Soleimani, A.; Pashirzad, M.; Abdeahad, H.; Mohammadi, F.; Khoshakhlagh, M.; Khazaei, M.; Ferns, G.A.; Avan, A.; Hassanian, S.M. Role of thrombin in the pathogenesis of atherosclerosis. *J. Cell. Biochem.* **2019**, *120*, 4757–4765. [CrossRef]
11. Li, Z.; Delaney, M.K.; O'Brien, K.A.; Du, X. Signaling during platelet adhesion and activation. *Arterioscler. Thromb. Vasc. Biol.* **2010**, *30*, 2341–2349. [CrossRef]
12. Tsopanoglou, N.E.; Maragoudakis, M.E. Thrombin's central role in angiogenesis and pathophysiological processes. *Eur. Cytokine Netw.* **2009**, *20*, 171–179.
13. Jennings, L.K. Mechanisms of platelet activation: Need for new strategies to protect against platelet-mediated atherothrombosis. *Thromb. Haemost.* **2009**, *101*, 248–257. [CrossRef]
14. Lordan, R.; O'Keeffe, E.; Dowling, D.; Mullally, M.; Heffernan, H.; Tsoupras, A.; Zabetakis, I. The in vitro antithrombotic properties of ale, lager, and stout beers. *Food Biosci.* **2019**, *28*, 83–88. [CrossRef]
15. Seefeldt, H.F.; Larsen, F.H.; Viereck, N.; Petersen, M.A.; Engelsen, S.B. Lipid composition and deposition during grain filling in intact barley (hordeum vulgare) mutant grains as studied by ^{1}H HR MAS NMR. *J. Cereal Sci.* **2011**, *54*, 442–449. [CrossRef]
16. Shewry, P.R. Chapter 7-minor components of the barley grain: Minerals, lipids, terpenoids, phenolics, and vitamins. In *Barley*, 2nd ed.; Shewry, P.R., Ullrich, S.E., Eds.; AACC International Press: St. Paul, MI, USA, 2014; pp. 169–192.
17. Cozzolino, D.; Degner, S. An overview on the role of lipids and fatty acids in barley grain and their products during beer brewing. *Food Res. Int.* **2016**, *81*, 114–121. [CrossRef]
18. Marion, D.; Dubreil, L.; Douliez, J.-P. Functionality of lipids and lipid-protein interactions in cereal-derived food products. *OCL* **2003**, *10*, 47–56. [CrossRef]
19. Olšovská, J.; Vrzal, T.; Štěrba, K.; Slabý, M.; Kubizniaková, P.; Čejka, P. The chemical profiling of fatty acids during the brewing process. *J. Sci. Food Agric.* **2019**, *99*, 1772–1779. [CrossRef]
20. Bamforth, C.W. Progress in brewing science and beer production. *Annu. Rev. Chem. Biomol. Eng.* **2017**, *8*, 161–176. [CrossRef]
21. Moonjai, N.; Verstrepen, K.J.; Shen, H.Y.; Derdelinckx, G.; Verachtert, H.; Delvaux, F.R. Linoleic acid supplementation of a cropped brewing lager strain: Effects on subsequent fermentation performance with serial repitching. *J. I. Brew.* **2003**, *109*, 262–272. [CrossRef]
22. Moonjai, N.; Verstrepen, K.J.; Delvaux, F.R.; Derdelinckx, G.; Verachtert, H. The effects of linoleic acid supplementation of cropped yeast on its subsequent fermentation performance and acetate ester synthesis. *J. Inst. Brew.* **2002**, *108*, 227–235. [CrossRef]
23. Gordon, R.; Power, A.; Chapman, J.; Chandra, S.; Cozzolino, D. A review on the source of lipids and their interactions during beer fermentation that affect beer quality. *Fermentation* **2018**, *4*, 89. [CrossRef]
24. Olajire, A.A. The brewing industry and environmental challenges. *J. Clean. Prod.* **2012**. [CrossRef]
25. Mussatto, S.; Dragone, G.; Roberto, I. Brewers' spent grain: Generation, characteristics and potential applications. *J. Cereal Sci.* **2006**, *43*, 1–14. [CrossRef]

26. Connolly, A.; Piggott, C.O.; FitzGerald, R.J. Characterisation of protein-rich isolates and antioxidative phenolic extracts from pale and black brewers' spent grain. *Int. J. Food Sci. Technol.* **2013**, *48*, 1670–1681. [CrossRef]
27. Niemi, P.; Tamminen, T.; Smeds, A.; Viljanen, K.; Ohra-aho, T.; Holopainen-Mantila, U.; Faulds, C.B.; Poutanen, K.; Buchert, J. Characterization of lipids and lignans in brewer's spent grain and its enzymatically extracted fraction. *J. Agric. Food Chem.* **2012**, *60*, 9910–9917. [CrossRef] [PubMed]
28. Dos Santos Mathias, T.R.; de Mello, P.P.M.; ervulo, E.F.C. Solid wastes in brewing process: A review. *J. Brew. Distill.* **2014**, *5*, 1–9.
29. Bligh, E.G.; Dyer, W.J. A rapid method of total lipid extraction and purification. *Can. J. Biochem. Physiol.* **1959**, *37*, 911–917. [CrossRef] [PubMed]
30. Tsoupras, A.; Zabetakis, I.; Lordan, R. Platelet aggregometry assay for evaluating the effects of platelet agonists and antiplatelet compounds on platelet function in vitro. *MethodsX* **2019**, *6*, 63–70. [CrossRef]
31. Tsoupras, A.; Lordan, R.; Shiels, K.; Saha, S.K.; Nasopoulou, C.; Zabetakis, I. In vitro antithrombotic properties of salmon (*salmo salar*) phospholipids in a novel food-grade extract. *Mar. Drugs* **2019**, *17*, 62. [CrossRef]
32. Bravi, E.; Marconi, O.; Sileoni, V.; Perretti, G. Determination of free fatty acids in beer. *Food Chem.* **2017**, *215*, 341–346. [CrossRef]
33. Bravi, E.; Benedetti, P.; Marconi, O.; Perretti, G. Determination of free fatty acids in beer wort. *Food Chem.* **2014**, *151*, 374–378. [CrossRef]
34. Bravi, E.; Marconi, O.; Perretti, G.; Fantozzi, P. Influence of barley variety and malting process on lipid content of malt. *Food Chem.* **2012**, *135*, 1112–1117. [CrossRef]
35. Evans, D.E.; Goldsmith, M.; Redd, K.S.; Nischwitz, R.; Lentini, A. Impact of mashing conditions on extract, its fermentability, and the levels of wort free amino nitrogen (fan), β-glucan, and lipids. *J. Am. Soc. Brew. Chem.* **2012**, *70*, 39–49. [CrossRef]
36. Michiu, D.; Socaci, S.A.; Jimborean, M.A.; Mudura, E.; Fărcaş, A.C.; Biriş-Dorhoi, S.E.; Tofană, M. Determination of volatile markers from magnum hops in beer by in-tube extraction—gas chromatography-mass spectrometry. *Anal. Lett.* **2018**, *51*, 2967–2980. [CrossRef]
37. Praet, T.; Van Opstaele, F.; De Causmaecker, B.; Aerts, G.; De Cooman, L. Heat-induced changes in the composition of varietal hop essential oils via wort boiling on a laboratory scale. *J. Am. Soc. Brew. Chem.* **2016**, *74*, 212–223. [CrossRef]
38. Yoon, M.-Y.; Cha, B.; Kim, J.-C. Recent trends in studies on botanical fungicides in agriculture. *Plant Pathol. J.* **2013**, *29*, 1–9. [CrossRef]
39. Bocquet, L.; Sahpaz, S.; Hilbert, J.L.; Rambaud, C.; Rivière, C. Humulus lupulus l., a very popular beer ingredient and medicinal plant: Overview of its phytochemistry, its bioactivity, and its biotechnology. *Phytochem. Rev.* **2018**, *17*, 1047–1090. [CrossRef]
40. Eyres, G.; Dufour, J.-P. 22-hop essential oil: Analysis, chemical composition and odor characteristics. In *Beer in Health and Disease Prevention*; Preedy, V.R., Ed.; Academic Press: San Diego, CA, USA, 2009; pp. 239–254.
41. Anness, B.J. Lipids of barley, malt and adjuncts. *J. I. Brew.* **1984**, *90*, 315–318. [CrossRef]
42. Blasi, F.; Montesano, D.; Simonetti, M.S.; Cossignani, L. A simple and rapid extraction method to evaluate the fatty acid composition and nutritional value of goji berry lipid. *Food Anal. Method.* **2017**, *10*, 970–979. [CrossRef]
43. Morrison, W.R.; Coventry, A.M. Extraction of lipids from cereal starches with hot aqueous alcohols. *Starch Stärke* **1985**, *37*, 83–87. [CrossRef]
44. Almaguer, C.; Schönberger, C.; Gastl, M.; Arendt, E.K.; Becker, T. Humulus lupulus—A story that begs to be told. A review. *J. I. Brew.* **2014**, *120*, 289–314. [CrossRef]
45. Pavithra, P.S.; Mehta, A.; Verma, R.S. Aromadendrene oxide 2, induces apoptosis in skin epidermoid cancer cells through ROS mediated mitochondrial pathway. *Life Sci.* **2018**, *197*, 19–29. [CrossRef]
46. Kennedy, G.L. Toxicity of adipic acid. *Drug Chem. Toxicol.* **2002**, *25*, 191–202. [CrossRef]
47. Lentz, M. The impact of simple phenolic compounds on beer aroma and flavor. *Fermentation* **2018**, *4*, 20. [CrossRef]
48. Shinohara, T.; Kubodera, S.; Yanagida, F. Distribution of phenolic yeasts and production of phenolic off-flavors in wine fermentation. *J. Biosci. Bioeng.* **2000**, *90*, 90–97. [CrossRef]

49. Varsha, K.K.; Devendra, L.; Shilpa, G.; Priya, S.; Pandey, A.; Nampoothiri, K.M. 2,4-di-tert-butyl phenol as the antifungal, antioxidant bioactive purified from a newly isolated lactococcus sp. *Int. J. Food Microbiol.* **2015**, *211*, 44–50. [CrossRef] [PubMed]
50. Choi, S.J.; Kim, J.K.; Kim, H.K.; Harris, K.; Kim, C.-J.; Park, G.G.; Park, C.-S.; Shin, D.-H. 2,4-di-tert-butylphenol from sweet potato protects against oxidative stress in pc12 cells and in mice. *J. Med. Food* **2013**, *16*, 977–983. [CrossRef]
51. Bijak, M.; Ziewiecki, R.; Saluk, J.; Ponczek, M.; Pawlaczyk, I.; Krotkiewski, H.; Wachowicz, B.; Nowak, P. Thrombin inhibitory activity of some polyphenolic compounds. *Med. Chem. Res.* **2014**, *23*, 2324–2337. [CrossRef] [PubMed]
52. Policegoudra, R.S.; Goswami, S.; Aradhya, S.M.; Chatterjee, S.; Datta, S.; Sivaswamy, R.; Chattopadhyay, P.; Singh, L. Bioactive constituents of homalomena aromatica essential oil and its antifungal activity against dermatophytes and yeasts. *J. Mycol. Med.* **2012**, *22*, 83–87. [CrossRef]
53. Fidyt, K.; Fiedorowicz, A.; Strządała, L.; Szumny, A. Β-caryophyllene and β-caryophyllene oxide-natural compounds of anticancer and analgesic properties. *Cancer Med.* **2016**, *5*, 3007–3017. [CrossRef]
54. Tognolini, M.; Barocelli, E.; Ballabeni, V.; Bruni, R.; Bianchi, A.; Chiavarini, M.; Impicciatore, M. Comparative screening of plant essential oils: Phenylpropanoid moiety as basic core for antiplatelet activity. *Life Sci.* **2006**, *78*, 1419–1432. [CrossRef]
55. Nasopoulou, C.; Karantonis, H.C.; Andriotis, M.; Demopoulos, C.A.; Zabetakis, I. Antibacterial and anti-PAF activity of lipid extracts from sea bass (Dicentrarchus labrax) and gilthead sea bream (Sparus aurata). *Food Chem.* **2008**, *111*, 433–438. [CrossRef]
56. Sioriki, E.; Smith, T.K.; Demopoulos, C.A.; Zabetakis, I. Structure and cardioprotective activities of polar lipids of olive pomace, olive pomace-enriched fish feed and olive pomace fed gilthead sea bream (*Sparus aurata*). *Food Res. Int.* **2016**, *83*, 143–151. [CrossRef]
57. Lordan, R.; Walsh, A.M.; Crispie, F.; Finnegan, L.; Cotter, P.D.; Zabetakis, I. The effect of ovine milk fermentation on the antithrombotic properties of polar lipids. *J. Funct. Foods* **2019**, *54*, 289–300. [CrossRef]
58. Demopoulos, C.A.; Karantonis, H.C.; Antonopoulou, S. Platelet-activating factor—A molecular link between atherosclerosis theories. *Eur. J. Lipid Sci. Technol.* **2003**, *105*, 705–716. [CrossRef]
59. Mori, T.A.; Beilin, L.J.; Burke, V.; Morris, J.; Ritchie, J. Interactions between dietary fat, fish, and fish oils and their effects on platelet function in men at risk of cardiovascular disease. *Arterioscler. Thromb. Vasc. Biol.* **1997**, *17*, 279–286. [CrossRef]
60. Yoshida, K.; Nagatoishi, S.; Kuroda, D.; Suzuki, N.; Murata, T.; Tsumoto, K. Phospholipid membrane fluidity alters ligand binding activity of a g protein-coupled receptor by shifting the conformational equilibrium. *Biochemistry* **2019**, *58*, 504–508. [CrossRef]

foods

MDPI

Article

Production and Refinement of Omega-3 Rich Oils from Processing By-Products of Farmed Fish Species

Vida Šimat [1,*], Jelena Vlahović [1,2,*], Barbara Soldo [3], Danijela Skroza [4], Ivica Ljubenkov [3] and Ivana Generalić Mekinić [4]

1 Department of Marine Studies, University of Split, Ruđera Boškovića 37, HR-21000 Split, Croatia
2 Sardina d.o.o., Ratac 1, HR-21410 Postira, Croatia
3 Department of Chemistry, Faculty of Science, University of Split, Ruđera Boškovića 33, HR-21000 Split, Croatia; barbara@pmfst.hr (B.S.); ivica.ljubenkov4@gmail.com (I.L.)
4 Department of Food Technology and Biotechnology, Faculty of Chemistry and Technology, University of Split, Ruđera Boškovića 35, HR-21000 Split, Croatia; danci@ktf-split.hr (D.S.); gene@ktf-split.hr (I.G.M.)
* Correspondence: vida@unist.hr (V.Š.); jelena.vlahovic@sardina.hr (J.V.); Tel.: +385-21-510-192 (V.Š.); +385-21-420-280 (J.V.)

Received: 22 March 2019; Accepted: 12 April 2019; Published: 16 April 2019

Abstract: In this study, the effect of a four-stage chemical refining process (degumming, neutralization, bleaching, deodorization) on the quality parameters, fatty acid composition and volatile compounds of crude oils produced from processing by-products of farmed fish species (tuna, seabass and gilthead seabream) was evaluated. The quality of the oils was compared to commercially available cod liver oil on the basis of free fatty acid, peroxide value, *p*-anisidine, total oxidation (TOTOX), thiobarbituric acid reactive species (TBARS), oxidative stability at 80, 100 and 120 °C, tocopherol content, and volatile components, while the fatty acid profile and the proportion of polyunsaturated fatty acids (PUFAs) were used as an indicator of the nutritional values of fish oils. Quality parameters of the studied oils and oil oxidative stability were enhanced with refining and were within the limits recommended for fish oils without the loss of PUFAs. In tuna by-product refined oils, the proportion of PUFAs was over 40%, with 30% of eicosapentaenoic and docosahexaenoic fatty acids. The volatile compounds of the oils were quantified (in mg/kg) and major components were 2,4-heptadienal, pentadecane, 2,4-decadienal, 2,4-nonadienal and dodecane. The use of aquaculture by-products as an alternative source for fish oil production could contribute to a more sustainable and profitable aquaculture production, providing economic benefits for the producers and setting new standards for a fish by-product disposal strategy.

Keywords: *Thunnus thynnus*; *Dicentrarhus labrax*; *Sparus aurata*; by-products; fish oil; chemical refining; fatty acid profile; volatile components

1. Introduction

The total aquaculture production of finfish is estimated at 54.1 million tonnes with a moderate annual growth rate of 5.8% with 37 world countries producing more farmed than wild-caught fish [1]. The Mediterranean and Black Sea areas have 62% of unsustainable fished stocks, particulary, the whitefish species. With the intent of making fisheries more productive and sustainable, mariculture (marine aquaculture) of seabass (*Dicentrarhus labrax*) and gilthead seabream (*Sparus aurata*) has grown exponentially, with these two species almost entirely farmed in the Mediterranean. The species with the highest commercial value in the Mediterranean is the Atlantic bluefin tuna (*Thunnus thynnus*) with capture-based mariculture (capture of fish from the wild and their rearing in sea cages for periods ranging between 3 months to 2 years) [2]. The growth of aquaculture also reflects on the generation of aquaculture by-products. In some countries such as Norway or Iceland, a large share of by-products is utilized by many different industries [3], in contrast to the Mediterranean region where this is not the practice. To our knowledge there is no available data for utilization of farmed tuna or seabass and gilthead seabream by-products for fish oil production or any other purpose that would benefit the producers economically and could serve as a reference for a fish by-product disposal strategy.

The many benefits that marine lipids contribute to our overall health are well known and widely documented. High levels of nutritionally valuable long-chain polyunsaturated fatty acids (PUFAs), mainly eicosapentaenoic acid (EPA) and docosahexaenoic acid (DHA) along with their physicochemical properties, make fish lipids unique and a promising economically valuable product. Although the current situation with fish oil production is stable, future projections show that available wild marine resources are becoming increasingly limited while the demand for PUFA continues to rise. This not only suggests, but stresses the need for exploration and exploitation of alternative sources [4]. Rest raw materials of the fish processing industry, also known as by-products, vary within species and processing methods used, but generally include parts that remain after edible parts of the fish is removed: Heads, viscera, frames (bones with attached flesh), skins, fins, trimmings, blood and others not utilized for human consumption [5,6]. Rest raw materials should not be considered waste or less valuable than the main product [5]. The high nutritive value of fish by-products is related to the content of valuable minerals, vitamins, protein and lipid fractions thus they are used in wide range of purposes, from production of fish cakes, pies and nuggets, gelatin, sauces and other products for human consumption to production of biodiesel [7]. Bioactive components such as peptides, enzymes or collagens that are used by pharmaceutical, biomedical and biotechnological industries can also be extracted from fish by-products [1,6]. With the growing interest and scientific knowledge within the field of by-product utilization and properties of certain components such as protein and oil that can be extracted from fish by-products, the low-value products such as mince or fish meal have been replaced by products of higher value [5,8]. Different fish species have specific processing yields and consequently, generate different proportions of by-products. In addition, the quality and chemical composition of marine by-products vary between species, season and catching ground as well as post-capture handling and processing. Although not directly used for human consumption, ingredients from by-products are used as feed in terrestrial livestock industries and aquaculture farming and indirectly contribute to human food production. The possible limitations to by-product utilization are related to lower quality of the fish meal and fish oil in comparison to products obtained from whole fish and elevated enzymatic activity that makes them highly perishable. Several studies showed that oil from by-products can be successfully extracted and refined to remove some undesirable compounds (moisture, pigments, free fatty acids, phospholipids, minerals, off-flavours, etc.) which affect oil stability, overall quality and consumers' acceptability, without the loss of PUFAs which enables its further application [9–13].

The aim of this research was to: (i) Produce and characterize crude oils from processing by-products obtained from farmed fish (tuna, seabass, and seabream); (ii) investigate the changes in fish oil characteristics during a four-stage refining process (degumming, neutralization, bleaching, deodorization) and differences in PUFAs composition between crude and refined oils; (iii) compare oils extracted from whole waste generated during tuna harvesting and tuna liver; (iv) report the characteristics of oil extracted from by-products obtained after filleting of farmed seabream and seabass; (v) compare the quality parameters and fatty acid composition of the obtained crude and refined oils with commercial cod liver oil in order to investigate the potential recycling of fish wastes for conversion into products of higher value.

2. Materials and Methods

2.1. General

All the standards compounds, solvents and reagents used in the study were obtained from Sigma-Aldrich (St. Louis, MO, USA), Fluka (Neu-Ulm, Germany), Merck (Darmstadt, Germany) and Kemika (Zagreb, Croatia). Spectrophotometric measurements were performed on a SPECORD 200 Plus (Analytik Jena AG, Jena, Germany). The analyses of fatty acid methyl esters (FAMEs) and volatile compounds were carried out by gas chromatograph (GC, model 3900; Varian Inc., Lake Forest, CA, USA) with flame-ionization detection (FID). The oil oxidative stability was determined using Rancimat model 743 (Metrohm, Herisau, Switzerland). The α-tocopherol content was determined by high-performance liquid chromatography (HPLC, all components of Series 200; Perkin Elmer, Waltham, MA, USA).

2.2. Raw Materials and Fish Oil Extraction

Farmed bluefin tuna (*Thunnus thynnus*) by-products were collected throughout December 2016 and January 2017 during harvesting and the evisceration process (done on board of fishing boat). By-products were separated into two groups: (i) Over 2000 kg of by-products (gills and gut content) were used for tuna by-product crude oil production; (ii) approximately 100 kg of tuna livers were collected (about 930 g each) and used for tuna liver crude oil production. During farming, tunas were fed with small pelagic fish such as sardines and herring and the average weight at harvesting was around 67 kg. Additionally, the farmed fish by-product crude oil was produced from approximately 1000 kg of heads, gills and gut content from seabream (*Sparus aurata*) and seabass (*Dicentrarhus labrax*) recovered from a local filleting plant. The fish were farmed at the same farm and fed with dry pellets (Efico plus, BioMar SAS, Nersac, France). Samples were 400–600 g at harvesting. Each oil was produced in two batches. Special attention was given to temperature regime during raw material manipulation. The crude oils were produced by grinding the raw materials in an industrial grinder (MG 250; Scansteel, Slagelse, Denmark), cooking at 95 °C for 12 min (cooking chamber model C2; Alfa Laval, Søborg, Denmark), and then pressed in a screw expeller and centrifuged at 4200 rpm using a decanter centrifuge type AC0303 CentryFish 1000 (Alfa Laval) which automatically separates dry matter (fishmeal), water and oil. With the interest of retaining the quality of crude oils, they were stored at <4 °C in dark bottles and refined within two days. Commercially available cod liver oil (oleum jecoris aselli; Kemig d.o.o., Zagreb, Croatia) was purchased from the local pharmacy and used as the control oil.

2.3. Oil Refinement Process

The refinement process, involving four major stages: Degumming (to separate phospholipids), neutralizing (to decrease acidity), bleaching (to remove coloured materials) and deodorization (to remove unwanted odour compounds), described by Chakraborty and Joseph [14] with slight modifications [15] was used. Extracted crude oil (500 g) was mixed with phosphoric acid (5 mL). The mixture was stirred in water bath heated to 70 °C for 20 min. After the cooling, samples were centrifuged (20 min, 4000 rpm) in order to remove the precipitated gum (degumming). Afterward, neutralization was done by slowly adding (drop by drop) NaOH solution (1 M) to the degummed oil samples in combination with constant stirring and heating at 65 °C for 20 min. The neutralization process was conducted until a pH of 7.0 was attained, and samples were then heated to 70 °C for 20 min, cooled and centrifuged for 15 min at 4000 rpm. The oil thus obtained was washed with deionized water (3-times with 10 mL) by agitation (500 rpm) and heating at 50 °C under vacuum. Neutralized oil samples were separated by centrifugation (10 min, 2500 rpm). The oil obtained after neutralization was bleached with adsorbents (4 g/100 g of oil) containing 1.13 g of activated carbon and 22.5 g of Fuller's earth. The oil samples were stirred using a magnetic stirrer at 40 °C for 40 min under N_2. After cooling, oil samples were again separated by centrifugation (30 min, 3500 rpm). Deodorization of the bleached oil samples was carried out by distillation under vacuum conditions. The mixture of the oil obtained after bleaching and deionized water (20 mL) was heated to 95–97 °C under the vacuum for one hour, with continual stirring. The refining process was repeated three times for each batch of oil.

2.4. Analytical Methodologies

American Oil Chemist' Society [16] methodologies were followed for oils chemical characterization as follows: Acidic value (AV, Method no. Ca 5a-40) expressed as percent of oleic acid, peroxide value (PV, Method no. Cd 8-53) expressed as milliequivalents of O_2/kg of oil, and anisidine values (*p*-AV, Method no. Cd 18-90) calculated as described in Chakraborty and Joseph [13]. The total oxidation (TOTOX) values were calculated as TOTOX = (2 × PV) + *p*-AV, while thiobarbituric acid reactive substances (TBARS) values were determined by the spectrophotometric assay described by Ke and Woyewoda [17]. The analyses were repeated five times.

The resistance of fish oil to auto-oxidation was determined at three different temperatures (80, 100 and 120 °C) by Rancimat method [18]. The fish oil (3 g) oxidative stability was tested at three different temperatures (80, 100 and 120 °C), while the airflow in all experiments was constant (20 L/h). The results were expressed as induction periods (IPs) which are the measure of oil stability or shelf-life, defined as the time in hours required to reach the end-point of oxidation.

2.5. Tocopherol Content

The tocopherol content was analyzed using HPLC system equipped with an autosampler, vacuum degasser, binary pump, fluorescent detector and the column oven on the Ultra Silica column (150 × 4.6 mm, 5 μm; Restek, Bellefonte, PA, USA) by a method previously described by Šimat et al. [15]. In short, an aliquot of 20 μL was obtained after dissolving 0.5 g of fish oil in 5 mL of hexane and injected into the chromatographic system. The following gradient elution program was applied using solvent A (hexane) and solvent B (isopropanol) at a flow rate of 0.9 mL/min: 3% B for the first 15 min, followed by an increase in solvent B to 80% through 5 min, after which the solvent ratio was maintained for 8 min. Over the next minute, the solvents returned to the initial conditions and this ratio of solvent was maintained for another 11 min to ensure column stabilization. The temperature of the column was held at 30 °C and detection observed by a fluorescent detector (excitation 290 nm and emission 330 nm). The compound was identified according to the retention time and quantified through the calibration curve of the standard. The analyses were performed in duplicate.

2.6. Fatty Acid Profile

The fatty acid methyl esters (FAMEs), prepared by dissolving oil samples (0.1 g) in heptane (2 mL) and by the addition of 2 M KOH in methanol (0.2 mL), were analyzed by gas chromatography with flame-ionisation detection (GC-FID) using capillary column RTX 2560 (100 m × 0.25 mm i.d., coating thickness 0.25 μm; Restek) by a method previously described in Šimat et al. [15]. In short, 1 μL of the heptane layer was injected into the chromatograph with a split ratio 1:100, with the temperature of injector et at 225 °C and of the detector at 240 °C. At a constant flow rate of 3 mL/min, helium was used as the carrier gas. The initial oven temperature was 140 °C, held for 5 min, raised to 240 °C at a rate of 4 °C/min and held at 240 °C for 20 min. FAMEs were identified by comparing to with standards (Supelco 37 Component FAME Mix; Sigma-Aldrich). The analyses were performed in duplicates and the results are expressed as percentages of methyl esters of individual fatty acids.

2.7. Analysis of Volatiles

The volatile compounds of oils were analyzed using headspace solid phase microextraction (HS-SPME) coupled with GC-FID. The samples were prepared using 2-cm long fibres of divinylbenzene-carboxen-polydimethylsiloxane (DVB-CAR-PDMS, thickness 50/30 μm) obtained from Supelco (Bellefonte, PA, USA). A total of 2 g of oil was placed into a 20 mL head space vial and equilibrated at 40 °C for 5 min, inserting the fibre into the headspace for adsorption at 40 °C for 30 min, then transferring the fibre to the injector port for desorption at 250 °C for 1 min.

The quantitative analyses and separation of volatile compounds of major fish oil volatile components were performed by GC-FID and CP-WAX 57 CB column (50 m × 0.25 mm i.d., coating thickness 0.2 μm; Varian). At a constant flow rate of 2 mL/min, helium was used as the carrier gas. The GC oven temperature was programmed at an initial 40 °C for 4 min, raised to 190 °C at a rate of 5 °C/min and kept constant for 11 min. The temperature was then raised up to 200 °C applying the same rate of 5 °C/min. The detector temperature was maintained at 250 °C. The total analysis time for each sample was 67 min. The analyses were performed in duplicates.

The investigated compounds were identified by the retention time of the corresponding analytical standard, while the quantification was made by external calibration. As the applied method did not ensure adequate separation of 2,4-heptadienal and pentadecane, results for these two compounds are presented as their sum while their concentration was calculated using the calibration curve obtained for 2,4-heptadienal. GC Workstation Version 6.41 chromatographic software (Varian) was used for data collection and calculation [19].

2.8. Statistical Analysis

The obtained results are expressed as mean values ± standard deviation. The means of detected parameters were analyzed for significance by analysis of variance (one-way ANOVA) using Statgraphics Centurion (StatPoint Technologies Inc., Warrenton, VA, USA). Differences were considered to be significant at $p < 0.05$.

3. Results and Discussion

3.1. Chemical Characterization and Oxidative Stability of Fish Oils

The results of the chemical and compositional quality characterization of oils from tuna by-products, tuna liver, seabass and gilthead seabream by-products during four refining stages and their comparison to cod liver oil (control) are presented in Table 1. Despite the high temperature used during the oil production (cooking at 95 °C for 12 min) the temperature was found to have a poor influence on oil quality parameters. During preliminary studies we inspected quality parameters in oils extracted at different temperatures (from 65–95 °C) and compared it to Bligh and Dyer extracts [20] and found no significant difference in oxidative parameters [21]. It has been previously found that oil extraction temperature is weakly linked to the oxidative quality of produced oil, but strongly affected by omega-3 content of the raw material, providing confirmation of our finding [22]. To control the quality of fish oil properties, which is very labile to hydrolytic spoilage and oxidative deterioration, numerous standards with variable acceptable levels have been established [23]. The express lipolysis and oxidation of fish oils are the results of high autolytic activity and high content of PUFAs in fish tissues. It is expected that this process is even more susceptible for fish by-products. For this reason, fish oils usually have high free fatty acid (FFA) content. In this study, the results of FFA values of crude oils were low confirming that short cooking periods during the oil extraction, even at higher temperatures, did not cause significant hydrolysis. Only in tuna liver oil the FFA values were over 3. The refinement process ensured an additional decrease of FFA by 3.7, 32 and 47% in tuna by-product, tuna liver and seabass/seabream oil, respectively. Among the three studied oils, seabass and gilthead seabream by-product oil had the lowest FFA values, even lower than the control. The allowable limit of FFAs value for crude fish oil is in the range of 1–7% of oleic acid, usually 2–5% [24], but the general recommendation is that FFA values of edible oils should be ≤3.0%. This is important since FFAs have an impact on the oil organoleptic properties as well as oil compositional quality [25], can act as pro-oxidants which initiate the oxidation mechanism [26] and high FFA values are problematic during omega-3 extraction and biodiesel production [27].

Primary oxidation of oil used to monitor hydroperoxides formation is determined by PV and should be ≤5 meq O_2/kg for fish oils intended for human consumption [28,29]. Despite the increase of the PVs for studied oils after degumming step, final values obtained for the refined oils were below the limit of 5 meq O_2/kg (Table 1) which opens the possibility of using these oils for human consumption.

The influences of the refining steps on oil oxidation status and products expressed as *p*-AV, TOTOX and TBARS are presented in Table 1. Among crude oils, *p*-AV and TOTOX values for tuna by-product oil were the highest, followed by those for tuna liver oil and seabass and gilthead seabream by-product oil. The processing steps of degumming, neutralizing and bleaching caused reduction of *p*-AV and TOTOX while these parameters increased after the deodorization step. The highest *p*-AV value was detected for refined tuna by-product oil (19.5), while TOTOX values of refined tuna liver and seabass and gilthead seabream by-product oils were higher than those obtained for crude oils, 27.7 and 24.7, respectively. This suggests that used adsorbents (activated carbon and Fuller's earth) have the capacity to adsorb primary and secondary oxidation compounds [14]. The allowable limit of *p*-AV for acceptability of fish oil for human consumption is ≤20 [30].

Although the *p*-AV obtained for oils in this study were under this limit, they were significantly higher than the control, thus additionally influencing the higher TOTOX value. The TOTOX is a parameter used to determine the presence of compounds generated by degradation of PUFAs under pro-oxidant conditions including high temperatures, oxygen, metal compounds and light, and the TOTOX value ≤ 26 under is found to be allowable for fish oil [31]. Therefore, in association to the above mentioned PV and *p*-AV values, the TOTOX values of studied by-product oils were above the mentioned limit in crude and refined oils from tuna by-products and just under the limit in seabass and seabream by-product oil (Table 1). The *p*-AV value and TOTOX obtained for crude sardine oil were found to be 16 and 40, respectively [14]. In that study, authors applied the same 4-stage refining process which was more effective for sardine oil and reduced *p*-AV value and TOTOX values to 10 and 19, respectively. On the other hand, as reported in Table 1, TBARS assay detected low accumulation of secondary oxidation products. TBARS detects lipid oxidation when thiobarbituric acid and oxidation products from unsaturated FAs react involving several secondary oxidation products, however, the TBARS of by-product oils was lower that the control oil.

Table 1. Chemical and quality parameters of crude oils from tuna by-products, tuna liver, seabass and gilthead seabream by-products at different stages of the refining process, and cod liver oil (control).

Oil Samples	Refining Phase	Measured Parameter				
		FFA (% Oleic Acid)	PV (meq O_2/kg)	*p*-AV	TOTOX	TBARS (μM/g)
Tuna by-product oil	*Crude oil*	2.94 ± 0.12 ^{a}	2.31 ± 0.73 ^{a}	25.2 ± 0.08 ^{a}	29.7 ± 0.20 ^{a}	0.94 ± 0.33 ^{a}
	Degummed	3.51 ± 0.03 ^{b}	3.93 ± 0.18 ^{b}	19.2 ± 1.25 ^{b}	27.1 ± 0.79 ^{b}	1.02 ± 0.14 ^{a}
	Neutralized	3.13 ± 0.10 ^{b}	3.46 ± 0.39 ^{b}	16.9 ± 1.23 ^{c}	23.8 ± 1.06 ^{c}	2.02 ± 0.45 ^{b}
	Bleached	2.93 ± 0.04 ^{a}	2.88 ± 0.11 ^{ab}	14.4 ± 0.90 ^{d}	20.2 ± 1.34 ^{d}	1.88 ± 0.37 ^{b}
	Deodorized	2.83 ± 0.01^{a}	3.78 ± 0.22 ^{b*}	19.5 ± 0.94 ^{b}	27.1 ± 2.15 ^{b}	1.14 ± 0.11 ^{a}
Tuna liver oil	*Crude oil*	3.13 ± 0.00 ^{a}	2.68 ± 0.86 ^{a}	20.9 ± 1.34 ^{a}	26.0 ± 0.14 ^{a}	2.88 ± 0.73 ^{a}
	Degummed	2.52 ± 0.05 ^{b}	3.01 ± 0.05 ^{ab}	19.1 ± 2.47 ^{b}	25.2 ± 1.17 ^{a}	1.35 ± 0.02 ^{b}
	Neutralized	2.65 ± 0.11 ^{c}	2.89 ± 0.22 ^{ab}	17.0 ± 4.78 ^{c}	22.7 ±1.22 ^{b}	1.76 ± 0.03 ^{b}
	Bleached	2.17 ± 0.03 ^{b}	2.85 ± 0.53 ^{ab}	14.0 ± 0.65 ^{d}	19.7 ± 0.43 ^{c}	1.92 ± 0.35 ^{b}
	Deodorized	2.12 ± 0.16 ^{b}	4.25 ± 0.28 ^{c}	19.2 ± 2.22 ^{b}	27.7 ± 2.22 ^{a}	1.97 ± 0.53 ^{b}
Seabass and gilthead seabream by-product oil	*Crude oil*	2.41 ± 0.47 ^{a}	4.63 ± 1.07 ^{a}	18.1 ± 0.19 ^{ab}	19.3 ± 2.45 ^{a}	0.53 ± 0.05 ^{a}
	Degummed	1.66 ± 0.05 ^{b}	2.70 ± 0.18 ^{b}	15.7 ± 1.12 ^{c}	20.2 ± 0.12 ^{a}	3.46 ± 2.08 ^{b}
	Neutralized	1.43 ± 0.18 ^{b}	3.13 ± 0.31 ^{b}	12.7 ± 0.65 ^{a}	18.9 ± 0.26 ^{a}	2.24 ± 1.05 ^{bc}
	Bleached	1.40 ± 0.01 ^{b}	3.44 ± 0.25 ^{ab}	11.4 ± 0.19 ^{b}	18.4 ± 0.52 ^{a}	1.14 ± 0.10 ^{ac}
	Deodorized	1.28 ± 0.15 ^{b}	4.20 ± 0.15 ^{ab}	16.5 ± 0.39 ^{c}	24.7 ± 0.32 ^{b}	1.42 ± 0.53 ^{ac*}
Cod liver oil	*Control*	2.15 ± 0.08	3.85 ± 0.44	13.3 ± 1.11	21.0 ± 0.18	1.63 ± 0.07

n = 30; Different superscript letters (a–d) in the same column denote statistically significant difference ($p < 0.05$);
* Measured parameters in deodorized oils marked with * do not differ statistically ($p < 0.05$) from the control sample.

The Rancimat method, an accelerated method that employs high temperatures and air-flow supply to estimate the oxidative stability and shelf life of oil-containing products in a relatively short time, was used to measure the oxidative stability of oils (Figure 1). According to the presented results it can be seen that tuna by-product oil was the most stabile sample among crude oils with IP of 1.54 h at 80 °C, 0.58 h at 100 °C and 0.15 h at 120 °C, while lower values were obtained for the other two oils. The refinement process prolonged the oil oxidative stability in all cases, but unlike for crude oils, refined tuna liver oil and seabass and seabream by-product oil showed higher IP values than tuna by-product oil (prolongation of 36% at 80 °C, 35% at 100 °C and 52 % at 120 °C). The oil resistance to the lipid oxidation is a result of differences between fatty acid profiles of investigated oils. From the Table 2 and Figure 2 it can be seen that crude tuna by product oil contain the highest content of PUFAs (∑ of 37.65%), as well as EPA + DHA (∑ of 30.85%). Furthermore, the *n-3/n-6* ratio in crude tuna waste oil was more than 3.5-fold higher than in crude tuna liver oil, and more than 9.5-fold higher than in crude seabass and seabream by-product oil. The refinement process reduced this parameter by half in tuna by-product oil, slight reduction was obtained in seabass and seabream by-product oil, while a higher value was detected in refined tuna liver oil. Although the referent cod liver oil showed the highest stability at 80 °C, at higher temperatures its stability was lower that of the investigated refined oils. As can be seen in Figure 3, significantly higher content of tocopherol in cod liver oil has been detected. This compound has been added during the production of the commercial oil sample in order to improve it stability, but it has been established that tocopherol degrades at higher temperatures [32] which is probably caused lower IP of the cod liver at 100 and 120 °C.

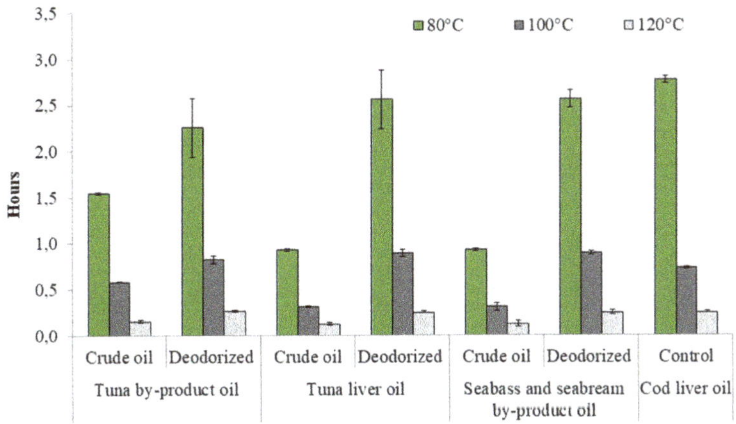

Figure 1. The oxidative stability (n = 18) of crude and deodorized (refined) oils from tuna by-products, tuna liver, seabass and gilthead seabream by-products, and cod liver oil (control).

3.2. Fatty Acid Profile of Oils

The results of the fatty acid profile of the crude oils and their changes after oil refinement are presented in Table 2 and in Figure 2A–D. The most dominant SFA contributing approximately 52–55% of total SFA, in all investigate oils was palmitic FA (C16:0). Its content was significantly higher in tuna oils in comparison to seabream/seabass oil and the control. Palmitic FA occurs naturally in fish, being a source of metabolic energy for their growth. The oleic acid (C18:1*n-9 cis*) was the major compound among MUFAs and in the studied oils amounted to approximately 14% in tuna oils, to 40% in seabass and gilthead seabream by-product refined oil. Among PUFAs, high concentrations of EPA (C20:5*n-3*) and even higher those of DHA (C22:6*n-3*) were found in all studied oils. The relative contents of EPA and DHA were expected to increase during the refining process [14,33], but in this study the minimal increase was observed after refinement only for DHA in tuna liver oil. However, in both crude and refined tuna by-product oils amounts of EPA and DHA were found to be extremely high. In tuna oils and the control oil, especially in tuna liver oil dihomo-γ-linolenic acid (20:3*n-6*), was found in higher amounts (7.32-9.89%), while in seabream/seabass oil its precursor, linoleic acid (C18:2*n-6*) was found significantly higher compared to other oils (17.32%). Erucic acid (22:1*n-9*) was found in small amounts in studied oils, with the highest content (1%) found in the control oil.

The crude oils from tuna by-products and liver contained significantly higher total amounts of saturated fatty acids (32.7 and 35.6%, Figure 2) in comparison to seabass and gilthead seabream by-product oil (23.6%). The SFA profile of the oils did not statistically change with the refining process and tuna oils had significantly higher SFA content than the control. The MUFAs content was the highest in seabass and gilthead seabream by-product oil (46%) and did not change after refining. Tuna by-product oil had slightly higher MUFA content then tuna liver oil. The PUFAs were significantly higher in tuna by-product and tuna liver crude oils (37.7 and 35.2%) and they increased to 40.2 and 39.6% in refined oils, respectively. The content of PUFA*n-3*, in the studied oils, ranged from 11.6–32.3% and was higher than that of PUFA*n-6* for 8.9-18.7% (Figure 2). A high percentage of PUFA indicates good nutritional values of the studied fish oils. The PUFA/SFA ratio of 0.4-0.5 is considered beneficial for human health [34], and it was significantly higher in the crude oils, while in refined oils it ranged from 1.1 to 1.2, significantly lower than in the control oil (Figure 2). In general, it is accepted that compared to wild caught fish, modern aquaculture products have lower *n-3* FAs, and higher levels of terrestrial plant-originating C18:2*n-6* as a result of feed composition [35]. Very high intake of *n-6* was recognized as undesirable and it reduces the nutritional quality of fish oil. The sum of n-6 was 18-19% higher in seabass and gilthead seabream by-product oils and along with the lowest sum of *n-3* and EPA+DHA.

Table 2. The changes in fatty acid profile (%) of crude and refined oils from tuna by-products, tuna liver, seabass and gilthead seabream by-products and comparison to cod liver oil (control).

Fatty Acid	Tuna By-Product Oil		Tuna Liver Oil		Seabass and Seabream By-Product Oil		Cod Liver Oil
	Crude Oil	*Refined*	*Crude Oil*	*Refined*	*Crude Oil*	*Refined*	*Control*
C12:0	n.d. [1]	0.06 ± 0.00	0.07 ± 0.00	0.06 ± 0.01	0.02 ± 0.01	0.03 ± 0.00	0.02 ± 0.02
C13:0	n.d.	0.06 ± 0.00	0.07 ± 0.00	0.07 ± 0.02	n.d.	0.02 ± 0.00	0.01 ± 0.01
C14:0	6.08 ± 0.98	6.05 ± 0.00	6.60 ± 0.49	6.50 ± 0.25	2.31 ± 0.02	2.57 ± 0.14	4.64 ± 0.02
C14:1	0.46 ± 0.65	0.04 ± 0.06	0.91 ± 0.06	n.d.	0.36 ± 0.00	0.40 ± 0.00	0.39 ± 0.03
C15:0	0.37 ± 0.52	0.82 ± 0.01	0.07 ± 0.00	0.88 ± 0.02	0.04 ± 0.01	0.05 ± 0.00	0.12 ± 0.01
C15:1	n.d.	0.01 ± 0.01	0.01 ± 0.00	n.d.	n.d.	n.d.	n.d.
C16:0	18.20 ± 2.12	18.09 ± 0.23	19.85 ± 1.54	18.97 ± 0.82	12.70 ± 0.02	13.87 ± 0.23	12.16 ± 0.18
C16:1	6.14 ± 0.74	6.09 ± 0.06	5.61 ± 0.43	5.40 ± 0.22	3.55 ± 0.09	3.88 ± 0.17	8.81 ± 0.01
C17:0	0.67 ± 0.07	0.65 ± 0.00	0.75 ± 0.06	0.71 ± 0.03	0.32 ± 0.02	0.34 ± 0.01	0.19 ± 0.02
C17:1	0.46 ± 0.10	0.44 ± 0.01	0.40 ± 0.04	0.35 ± 0.02	0.27 ± 0.01	0.29 ± 0.01	0.35 ± 0.02
C18:0	4.28 ± 0.36	4.23 ± 0.01	4.97 ± 0.36	4.66 ± 0.08	2.63 ± 0.01	0.17 ± 0.00	3.21 ± 0.34
C18:1n-9t	2.64 ± 0.41	0.16 ± 0.00	0.17 ± 0.02	0.16 ± 0.01	0.12 ± 0.00	0.15 ± 0.00	0.14 ± 0.00
C18:1n-9c	13.92 ± 1.14	13.73 ± 0.18	13.42 ± 1.09	13.05 ± 0.09	37.97 ± 0.08	40.52 ± 0.72	18.73 ± 0.01
C18:2n-6t	0.06 ± 0.08	0.05 ± 0.03	0.08 ± 0.09	0.13 ± 0.01	0.04 ± 0.00	0.04 ± 0.00	0.08 ± 0.00
C18:2n-6c	1.04 ± 1.47	2.18 ± 0.01	2.28 ± 0.18	2.13 ± 0.02	16.35 ± 0.02	17.32 ± 0.26	1.89 ± 0.02
C20:0	0.42 ± 0.03	0.42 ± 0.00	0.47 ± 0.04	0.43 ± 0.01	0.51 ± 0.00	0.39 ± 0.01	0.29 ± 0.01
C18:3n-6	0.39 ± 0.55	0.13 ± 0.00	0.12 ± 0.01	0.11 ± 0.02	0.19 ± 0.00	0.19 ± 0.00	0.15 ± 0.00
C20:1	5.07 ± 0.20	5.08 ± 0.02	6.50 ± 0.56	5.84 ± 0.04	2.47 ± 0.00	2.50 ± 0.05	12.67 ± 0.04
C18:3n-3	1.41 ± 0.12	n.d.	n.d.	0.65 ± 0.06	0.02 ± 0.02	0.18 ± 0.00	n.d.
C21:0	n.d.	1.44 ± 0.04	1.48 ± 0.13	0.67 ± 0.08	4.28 ± 0.01	4.46 ± 0.05	0.90 ± 0.02
C20:2	0.28 ± 0.01	0.27 ± 0.01	0.31 ± 0.03	0.11 ± 0.01	0.59 ± 0.01	0.58 ± 0.02	0.30 ± 0.02
C22:0	0.08 ± 0.11	0.15 ± 0.01	0.18 ± 0.02	0.17 ± 0.02	0.19 ± 0.02	0.17 ± 0.01	0.14 ± 0.00
C20:3n-6	3.54 ± 5.01	7.32 ± 0.02	10.25 ± 0.92	9.04 ± 0.17	1.04 ± 0.00	0.97 ± 0.01	9.89 ± 0.02
C22:1n-9	0.32 ± 0.01	0.30 ± 0.00	0.39 ± 0.03	0.37 ± 0.01	0.46 ± 0.03	0.43 ± 0.01	1.00 ± 0.02
C20:3n-3	n.d.	0.02 ± 0.00	0.05 ± 0.00	0.11 ± 0.11	0.03 ± 0.00	0.02 ± 0.00	n.d.
C20:4n-6	0.19 ± 0.13	0.18 ± 0.00	0.19 ± 0.01	0.09 ± 0.14	0.16 ± 0.00	0.16 ± 0.00	0.16 ± 0.01
C23:0	1.02 ± 0.08	0.97 ± 0.00	1.01 ± 0.07	1.92 ± 0.01	0.51 ± 0.01	0.48 ± 0.01	0.58 ± 0.01
C22:2	n.d.	0.03 ± 0.00	0.05 ± 0.01	0.36 ± 0.07	0.05 ± 0.01	0.05 ± 0.00	0.02 ± 0.01
C24:0	0.87 ± 0.04	0.02 ± 0.00	0.07 ± 0.01	0.40 ± 0.02	0.07 ± 0.00	0.05 ± 0.01	0.82 ± 0.04
C20:5n-3	9.56 ± 0.76	9.29 ± 0.04	8.81 ± 0.68	8.21 ± 0.03	3.33 ± 0.03	3.04 ± 0.00	9.58 ± 0.06
C24:1	0.60 ± 0.02	0.65 ± 0.00	0.90 ± 0.09	0.77 ± 0.14	0.45 ± 0.01	0.34 ± 0.04	0.48 ± 0.00
C22:6n-3	21.29 ± 1.09	20.75 ± 0.19	13.07 ± 7.00	18.66 ± 1.21	8.26 ± 0.24	6.37 ± 0.47	11.9 ± 0.26

[1] n.d.—not detected; $n = 12$.

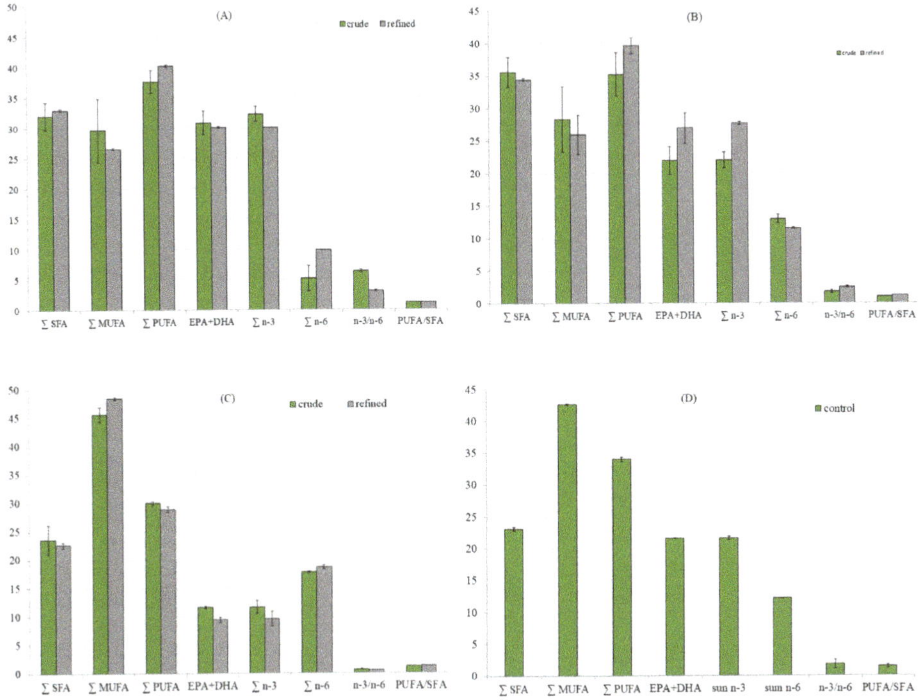

Figure 2. The sum of monounsaturated (∑MUFA) and polyunsaturated (∑PUFA) fatty acids, the sum of eicosapentaenoic and docosahexaenoic fatty acids (EPA + DHA), the sum of *n-3* and *n-6* content plus ratios between *n-6/n-3* and polyunsaturated fatty acids/saturated fatty acids (PUFA/SFA) in crude and refined oils from: (**A**) tuna by-products, (**B**) tuna liver, (**C**) seabass and gilthead seabream by-products and (**D**) cod liver oil (control).

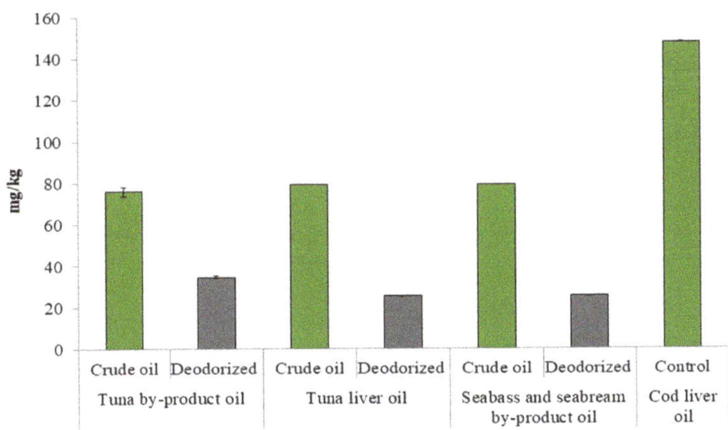

Figure 3. The tocopherol content ($n = 12$) in crude and deodorized (refined) oils from tuna by-products, tuna liver, seabass and gilthead seabream by-products and cod liver oil (control).

In tuna by-product oils the refining process resulted with a decrease share of MUFAs and the increase of PUFAs. This, as well as the content of antioxidant compounds such as tocopherol (Figure 3), can contribute to the oxidative stability of oils. Crude oils from farmed fish by-products had high tocopherol content which was significantly reduced by refining process in all oils (31–45%). At the same time, oxidative stability was prolonged by the refining process thus the role of tocopherol in oxidative stability appears to be smaller than the removal impurities which act as pro-oxidants [36]. This is confirmed on the control oil which had over 140 mg of tocopherol per kilogram (data from declaration sheet indicate a content of 1190 International Units of vitamin A per gram) however this content did not enhance its oxidative stability at elevated temperatures. The tocopherol content of oils from farmed fish species is higher in comparison to oils from wild fish and their by-products. For example, sardine by-product crude oil has approximately 30 mg tocopherol /kg [15]. It has been suggested that dietary elements of fish feed, such as vitamin E, do not influence significantly the amount of total lipids, phospholipids, polyunsaturated and general muscle fatty acid composition but protect from peroxide formation and phospholipid hydrolysis [35].

3.3. Volatile Profile of Oils

The composition and relative contents of volatile compounds of crude, bleached and deodorized oils from tuna by-products, tuna liver, seabass and gilthead seabream by-products and cod liver oil (control) are presented in Table 3. Sixteen volatile components were identified (one ester, six aldehydes, five alcohols and three hydrocarbons). In tuna by-product oil the most dominant was the sum of 2,4-heptadienal (from *n-3* fatty acids) and pentadecane, followed by (E,E)-2,4-decadienal, dodecane and 4-methylpenten-2-ol. The applied GC-FID method did not ensure adequate separation of 2,4-heptadienal and pentadecane, and their concentration was calculated using the calibration curve obtained for 2,4-heptadienal since it was considered more important in fish oil, however the exact amount of the 2,4-heptadienal in the total sum is unknown. Taking into account that other volatile compounds such as (E,Z)-2,6-nonadienal and (E,E)-2,4-decadienal (secondary lipid oxidation products) were found in low concentrations, low TBARs values (Table 1) and previous reports that suggest pentadecane as dominant component in fish oil samples [37,38], we can assume that pentadecane is the dominant compound in this mixture. Similar was observed for tuna liver oil with exception to high levels of 1-penten-3-ol. Among unsaturated alcohols responsible for the fishy odour of the oil, one of the most important components is 1-pentene-3-ol [39]. The content of this compound was significantly reduced in all studied oils and in refined oils, with the findings ranging from 0.01–1.03 mg/kg.

In seabass and gilthead seabream by-products oil aldehydes, (E,E)-2,4-decadienal and 2,4-nonadienal, were found to be higher than in tuna by-product oils and levels of 2,4-heptadienal+pentadecane were significantly lower. The amount of 2,4-decadienal (from the *n-6* fatty acids) increased after the distillation, especially in seabass and seabream oil. In comparison to studied oils, the control oil was characterized with high levels of tetradecene. The values of two fatty aldehydes, 2,4-heptadienal and 2,4-decadienal are of great importance due to their contribution to the characteristic unpleasant odour of the oil [14,39]. The composition and proportion of volatile compounds changed significantly during the refining process and only in tuna by-product oil the total sum of volatile compounds was reduced during refining. The aldehydes are known as essential indicators of the oxidation processes and are also responsible for oil fishy odour, same for ketones which usually have very low thresholds and are derived from autoxidation of PUFAs via hydroperoxides or lipid oxidative degradation. In order to remove undesirable favor components, such as oxidation products (aldehydes and ketone, residual free fatty acids, etc.) the deodorization step is generally carried out by conventional steam distillation at temperatures below 200 °C. The suggested procedure does not affect all volatile components equally. The effectiveness of this process is influenced by the applied pressure and volatility of components at high temperature. In this stage, it is also important to inhibit the degradation of the essential components by cyclization and polymerization of long chain PUFAs [26].

Table 3. The concentration of volatiles (mg/kg) in crude oils from tuna by-products, tuna liver, seabass and gilthead seabream by-products at different stages of the refining process, and cod liver oil (control).

	Volatile Compound	Tuna By-Product Oil			Tuna Liver Oil			Seabass and Seabream By-Product Oil			Cod Liver Oil
		Crude	*Bleached*	*Deodorized*	*Crude*	*Bleached*	*Deodorized*	*Crude*	*Bleached*	*Deodorized*	*Control*
Esters	Ethyl acetate	1.51 ± 0.14 ^{a}	1.45 ± 0.19 ^{a}	1.00 ± 0.02 ^{b*}	0.93 ± 0.28 ^{a}	1.03 ± 0.29 ^{a}	1.19 ± 0.01 ^{a*}	0.50 ± 0.03 ^{a}	1.69 ± 0.25 ^{b}	0.99 ± 0.07 ^{c*}	0.73 ± 0.07
Aldehydes	Pentanal	1.12 ± 0.14 ^{a}	1.55 ± 0.06 ^{a}	1.94 ± 0.32 ^{a*}	0.95 ± 0.20 ^{ab}	0.81 ± 0.05 ^{ab}	0.54 ± 0.01 ^{bc}	2.09 ± 0.05 ^{a}	1.53 ± 0.11 ^{b}	0.55 ± 0.04 ^{c}	1.22 ± 0.11
	E-2-hexenal	0.08 ± 0.01 ^{a}	0.05 ± 0.01 ^{a}	0.05 ± 0.02 ^{a}	0.04 ± 0.01 ^{a}	0.04 ± 0.01 ^{a}	0.02 ± 0.0 ^{a}	0.01 ± 0.0 ^{a}	0.01 ± 0.0 ^{a}	0.02 ± 0.0 ^{a}	0.51 ± 0.04
	Octanal	0.29 ± 0.25 ^{a}	0.53 ± 0.0 ^{a}	0.39 ± 0.24 ^{a*}	1.15 ± 0.08 ^{a}	0.41 ± 0.20 ^{a}	0.45 ± 0.10 ^{a*}	0.34 ± 0.13 ^{ab}	0.31 ± 0.02 ^{ab}	0.15 ± 0.09 ^{a*}	0.14 ± 0.02
	(E,Z)-2,6-nonadienal	0.98 ± 0.21 ^{a}	1.07 ± 0.14 ^{a}	0.85 ± 0.13 ^{a}	0.91 ± 0.01 ^{a}	1.20 ± 0.16 ^{a}	1.08 ± 0.08 ^{a}	0.40 ± 0.06 ^{a}	0.62 ± 0.09 ^{b}	0.65 ± 0.03 ^{b}	0.14 ± 0.0
	(E,E)-2,4-decadienal	4.59 ± 0.28 ^{a}	6.06 ± 0.57 ^{b}	4.94 ± 0.35 ^{ab}	6.28 ± 1.19 ^{a}	6.25 ± 0.70 ^{a}	6.76 ± 0.42 ^{a*}	11.5 ± 0.46 ^{a}	10.3 ± 0.18 ^{a}	14.1 ± 0.54 ^{b}	6.80 ± 0.21
	2,4-nonadienal	0.67 ± 0.04 ^{a}	1.03 ± 0.14 ^{a}	1.02 ± 0.13 ^{a}	2.09 ± 0.16 ^{a}	0.93 ± 0.13 ^{b}	0.96 ± 0.13 ^{b}	6.53 ± 0.58 ^{a}	6.59 ± 0.49 ^{a}	4.77 ± 0.02 ^{b}	0.26 ± 0.05
Alcohols	1-penten-3-ol	0.08 ± 0.0 ^{a}	0.05 ± 0.01 ^{ab}	0.01 ± 0.01 ^{b*}	11.17 ± 0.20 ^{a}	2.29 ± 1.31 ^{b}	1.03 ± 0.0 ^{b}	0.65 ± 0.01 ^{a}	0.37 ± 0.02 ^{b}	0.07 ± 0.0 ^{c}	0.02 ± 0.0
	4-methylpenten-2-ol	2.66 ± 0.07 ^{a}	1.42 ± 0.17 ^{b}	2.15 ± 0.57 ^{a}	1.14 ± 0.03 ^{a}	1.08 ± 0.57 ^{a}	1.46 ± 0.0 ^{a*}	0.57 ± 0.03 ^{a}	1.55 ± 0.10 ^{b}	1.91 ± 0.10 ^{c}	1.14 ± 0.03
	Hexanol	0.08 ± 0.01 ^{a}	0.03 ± 0.01 ^{b}	0.05 ± 0.03 ^{b*}	0.06 ± 0.0 ^{a}	0.03 ± 0.01 ^{a}	0.04 ± 0.0 ^{a*}	0.06 ± 0.0 ^{a}	0.02 ± 0.0 ^{b}	0.03 ± 0.0 ^{c}	0.06 ± 0.01
	E-2-hexen-1-ol	0.08 ± 0.0 ^{a}	0.06 ± 0.04 ^{ab}	0.11 ± 0.03 ^{a}	0.01 ± 0.01 ^{a}	0.13 ± 0.09 ^{a}	0.23 ± 0.02 ^{a}	0.04 ± 0.0 ^{a}	0.12 ± 0.01 ^{b}	0.23 ± 0.02 ^{c}	n.d.
	Z-3-hexen-1-ol	0.05 ± 0.01 ^{ab}	0.04 ± 0.01 ^{ab}	0.03 ± 0.0 ^{a}	0.05 ± 0.01 ^{a}	0.20 ± 0.14 ^{a}	0.06 ± 0.0 ^{a*}	0.01 ± 0.01 ^{a}	0.06 ± 0.0 ^{b}	0.02 ± 0.0 ^{a}	0.11 ± 0.01
Hydrocarbons	Dodecane	3.85 ± 0.58 ^{a}	4.03 ± 0.11 ^{a}	3.50 ± 0.55 ^{a}	3.22 ± 0.15 ^{a}	4.41 ± 0.59 ^{a}	3.23 ± 0.74 ^{a}	1.17 ± 0.11 ^{a}	1.79 ± 0.59 ^{a}	1.39 ± 0.08 ^{a}	0.76 ± 0.02
	Tetradecane	1.13 ± 0.22 ^{a}	1.19 ± 0.26 ^{a}	0.71 ± 0.41 ^{a*}	0.85 ± 0.01 ^{a}	0.60 ± 0.07 ^{b}	1.09 ± 0.10 ^{c}	1.18 ± 0.06 ^{a}	1.00 ± 0.06 ^{a}	1.09 ± 0.15 ^{a}	0.51 ± 0.01
	Tetradecene	1.22 ± 0.02 ^{a}	1.17 ± 0.21 ^{a}	1.09 ± 0.15 ^{a}	1.32 ± 0.19 ^{a}	1.18 ± 0.08 ^{a}	1.02 ± 0.10 ^{a}	2.66 ± 0.16 ^{a}	2.49 ± 0.37 ^{a}	2.31 ± 0.07 ^{a}	8.62 ± 0.52
	2,4-heptadienal +pentadecan	174 ± 22.6 ^{a}	208 ± 27.5 ^{b}	160 ± 19.6 ^{a}	146 ± 11.5 ^{a}	164 ± 10.5 ^{a}	172 ± 9.43 ^{a}	36.0 ± 4.65 ^{a}	51.4±8.07 ^{a}	86.5 ± 0.52 ^{b}	11.0 ± 0.11

$^{a-c}$ Different subscript letters define statistically significant difference (p < 0.05) among different stages of the refining process for each oil; * Concentration of volatiles in deodorized oils marked with * do not differ statistically (p < 0.05) from control sample; Deodorized = refined; n.d.—not detected; n = 12.

Chakraborty and Joseph [14] reported that sardine fish oil distillate obtained after 60 min of steam distillation under vacuum contained two prominent aldehydes, namely, 2,4-heptadienal and 2,4-decadienal. The concentrations of these compounds were also high in our study suggesting that improvement of the distillation method is necessary.

The aroma components of fish have been widely studied, however volatile compound profiles in crude fish by-products oils and in oils undergoing the refining process have rarely been reported. In general, volatile compounds of fish oils are only identified [14,33,40], without dealing with the quantification of those components (concentrations reported in mg/kg instead of in % of peak area). The volatile components in fish oils are usually a result of microbiological spoilage or oxidation processes of lipids, amino acids and proteins and the knowledge of their chemical characterization and changes during treatments is useful for feed and aquatic industry as it opens new usage possibilities of different processing by-products [14,33,41].

The changes of volatile profiles of fish oil, from tuna and anchovy by-products in chemical refining process have been previously reported by Song et al. [33]. Authors identified 63 volatile compounds, with hexanal, nonanal, undecanal, 2-nonanone, and 2-undecanone being the key volatile components of the fish oils. The study demonstrated that compounds which are most responsible for the unfavorable odour of the fish oil could be effectively removed by the refining process which directly enhances the oil quality. Oliveira et al. [42] studied the effects of chemical refining and deodorization on fatty acid profiles and sensory characteristics of the tuna (*Thunnus albacares*) by-product oil obtained by enzymatic hydrolysis. The oil was extracted from the heads and was found rich in PUFAs. In comparison to this study, authors found higher content of MUFAs in refined oil (36.78%) and lower content of PUFA (33.18%) and recommended deodorization conditions at 160 °C for 1 h and 200 °C for 1 h for PUFA rich oils.

4. Conclusions

The chemical and compositional quality of the characterization of oils from tuna by-products, tuna liver, seabass, and gilthead seabream by-products during the four stage refining process suggests that by-products are suitable and valuable raw material for PUFA-rich oil production. Crude oils showed good characteristics; and the four-stage refining process was effective in reducing oil impurities resulting in lower FFA, PV, *p*-AV, TOTOX, and TBARS values of studied oils. Refined oils also showed better oxidative stability. A beneficial compound, α-tocopherol, which might improve the oil quality was lowered by the described refining process in all oils. Volatile compounds responsible for the sensory profile of the oils were formed and removed from the oils during different refining steps without clear relation to FFA and TBARS values. In order to maintain high levels of long-chain fatty acids, such as EPA and DHA, the described refining procedure was effective, but for effective removal of volatile components from fish oil, improvement of the deodorization step is needed possibly at temperatures higher than 100 °C for a short period of time. During the search of an alternative source of raw material for fish oil production, the results of this paper have been found to contribute to a more efficient utilization of natural marine resources, clearly indicating that by-products may offer a solution that will improve waste management, an ecological aspect of fish processing, thereby adding value to "waste".

Author Contributions: Conceptualization, V.Š., J.V. and I.G.M.; methodology, V.Š., J.V., D.S., I.L.; formal analysis, J.V., D.S., B.S., I.L.; investigation, V.Š., J.V., D.S., B.S.; writing—original draft preparation, V.Š. and J.V.; writing—review and editing, I.G.M. and D.S.; visualization, V.Š.

Funding: This research was funded by the Croatian Science Foundation grant number IP-2014-09-6897.

Acknowledgments: The authors would like to thank Sardina d.o.o. (Postira, Croatia) for raw material support, technical and facilities support of this research. We thank Heidi-Christine Santoro for the language editing.

Conflicts of Interest: The authors declare no conflict of interest.

References

1. FAO. *The State of World Fisheries and Aquaculture 2018—Meeting the Sustainable Development Goals*; FAO: Rome, Italy, 2018; p. 227.
2. Constantinos, C.M.; De La Gándara, F.; Corriero, A.; Belmonte Ríos, A. Atlantic bluefin tuna (*Thunnus Thynnus*) farming and fattening in the Mediterranean Sea. *Rev. Fish. Sci.* **2010**, *18*, 266–280. [CrossRef]
3. Rustad, T. Utilization of marine by-product. *EJEAFChe* **2003**, *2*, 458–463.
4. Ivanovs, K.; Blumberga, D. Extraction of fish oil using green extraction methods: A short review. *Energy Procedia* **2017**, *128*, 477–483. [CrossRef]
5. Arason, S.; Karlsdottir, M.; Valsdottir, T.; Slizyte, R.; Rustad, T.; Falch, E.; Eysturskard, J.; Jakobsen, G. *Maximum Resource Utilisation—Value Added Fish By-Products*; Nordic Innovation Centre: Oslo, Norway, 2009; p. 108. Available online: http://nordicinnovation.org/Global/_Publications/Reports/2010/Maximum%20resource%20utilisation%20-%20Value%20added%20fish%20by-products.pdf (accessed on 24 January 2019).
6. Stevens, J.R.; Newton, R.W.; Tlusty, M.; Little, D.C. The rise of aquaculture by-products: Increasing food production, value, and sustainability through strategic utilisation. *Mar. Policy* **2018**, *90*, 115–124. [CrossRef]
7. Kara, K.; Ouanji, F.; El Lotfia, M.; El Mahia, M.; Kacimi, M.; Ziyad, M. Biodiesel production from waste fish oil with high free fatty acid content from Moroccan fish-processing industries. *Egypt. J. Pet.* **2017**, *27*, 249–255. [CrossRef]
8. Rustad, T.; Storro, I.; Slizyte, R. Possibilities for the utilisation of marine by-products. *Int. J. Food Sci. Technol.* **2011**, *46*, 2001–2014. [CrossRef]
9. Menegazzo, M.L.; Petenuci, M.E.; Fonseca, G.G. Production and characterization of crude and refined oils obtained from the co-products of Nile tilapia and hybrid sorubim processing. *Food Chem.* **2014**, *157*, 100–104. [CrossRef] [PubMed]
10. Huang, J.; Sathivel, S. Purifying salmon oil using adsorption, neutralization, and a combined neutralization and adsorption process. *J. Food Eng.* **2010**, *96*, 51–58. [CrossRef]
11. Ferdosh, S.; Sarker, Z.I.; Norulaini, N.; Oliviera, A.; Yunus, K.; Choedury, A.J.; Arkanda, J.; Omar, M. Quality of tuna fish oils extracted from processing the by- products of three species of neritic tuna using supercritical carbon dioxide. *J. Food Process Preserv.* **2015**, *39*, 432–441. [CrossRef]
12. Kuo, C.H.; Liao, H.Z.; Wang, Y.H.; Wang, H.M.D.; Shieh, C.J.; Tseng, C.Y. Highly efficient extraction of EPA/DHA-enriched oil from cobia liver using homogenization plus sonication. *Eur. J. Lipid Sci. Technol.* **2017**, *119*, 1600466. [CrossRef]
13. Nazir, N.; Diana, A.; Sayuti, K. Physicochemical and fatty acid profile of fish oil from head of tuna (Thunnus albacares) extracted from various extraction method. *IJASEIT* **2017**, *7*, 709–715. [CrossRef]
14. Chakraborty, K.; Joseph, D. Production and characterization of refined oils obtained from Indian oil sardine (*Sardinella longiceps*). *J. Agric. Food Chem.* **2015**, *63*, 998–1009. [CrossRef]
15. Šimat, V.; Vlahović, J.; Soldo, B.; Generalić Mekinić, I.; Čagalj, M.; Hamed, I.; Skroza, D. The quality of crude marine oils from seafood processing by-products: Production and characterization. *Food Biosci.* **2019**, in press.
16. AOCS. *The Official Methods and Recommended Practices of the American Oil Chemists' Society*; The American Oil Chemists' Society: Champaign, IL, USA, 1994.
17. Ke, P.; Woyewoda, A. Microdetermination of thiobarbituric acid values in marine lipids by a direct spectrophotometric method with a monophasic reaction system. *Anal. Chim. Acta* **1979**, *106*, 279–284. [CrossRef]
18. Šimat, V.; Ficović, M.; Čagalj, M.; Skroza, D.; Ljubenkov, I.; Generalić Mekinić, I. Preventive effect of herb extracts on lipid oxidation in fish oil. *Croat. J. Food Technol. Biotechnol. Nutr.* **2017**, *12*, 30–36.
19. Soldo, B.; Šimat, V.; Vlahović, J.; Skroza, D.; Ljubenkov, I.; Generalić Mekinić, I. Sardine by-products as an alternative to whole sardines in oil production: Chemical composition and quality parameters during refining process. *Eur. J. Lipid Sci. Technol.* **2019**, in press.
20. Bligh, E.G.; Dyer, W.J. A rapid method of total lipid extraction and purification. *Can. J. Biochem. Physiol.* **1959**, *37*, 911–917. [PubMed]
21. Šimat, V.; Vlahović, J. University Department of Marine Studies, University of Split, Split, Croatia. Unpublished work, 2016.

22. Honold, P.J.; Nouard, M.-L.; Jacobsen, C. Fish oil extracted from fish-fillet by-products is weakly linked to the extraction temperature but strongly linked to the omega-3 content of the raw material. *Eur. J. Lipid Sci. Technol.* **2016**, *118*, 874–884. [CrossRef]
23. Ceriani, R.; Paiva, F.R.; Goncalves, C.B.; Batista, E.A.C.; Meirelles, A.J.A. Densities and viscosities of vegetable oils of nutritional value. *Chem. Eng. Data* **2008**, *53*, 1846–1853.
24. Bimbo, A.P. Guidelines for characterization food-grade fish oil. *Inform* **1998**, *9*, 473–483.
25. Özyurt, G.; Şimşek, A.; Etyemez, M.; Polat, A. Fatty acid composition and oxidative stability of fish oil products in Turkish retail market. *J. Aquat. Food Prod. Technol.* **2013**, *22*, 322–329. [CrossRef]
26. Vaisali, C.; Charanyaa, S.; Belur, P.D.; Regupathi, I. Refining of edible oils: A critical appraisal of current and potential technologies. *Int. J. Food Sci. Technol.* **2015**, *50*, 13–23. [CrossRef]
27. Bako, T.; Victor, U.I.; Awulu, O. Criteria for the extraction of fish oil. *Agric. Eng. Int. CIGR J.* **2017**, *19*, 120–132.
28. FAO/WHO. Proposed draft standard for fish oils. In Proceedings of the 23rd Session Codex Committee on Fats and Oils, Langkawi, Malaysia, 25 February–1 March 2013.
29. Codex Alimentarius Commision. *Standards for Fish Oils*; CXS 329-2017 1–6; Codex Alimentarius Commision: Rome, Italy, 2017.
30. Crexi, V.T.; Monte, M.L.; Soares, L.A.S.; Pinto, L.A.A. Production and refinement of oil from carp (*Cyprinus carpio*) viscera. *Food Chem.* **2010**, *119*, 945–950. [CrossRef]
31. Deepika, D.; Vegneshwaran, V.R.; Julia, P.; Sukhinder, K.C.; Sheila, T.; Heather, M.; Wade, M. Investigation on oil extraction methods and its influence on omega-3 content from cultured salmon. *J. Food Process Technol.* **2014**, *5*, 12. [CrossRef]
32. Sabliov, C.M.; Fronczek, C.; Astete, C.E.; Khachaturyan, M.; Khachatryan, L.; Leonardi, C. Effects of temperature and UV light on degradation of α-tocopherol in free and dissolved form. *J. Am. Oil Chem. Soc.* **2009**, *86*, 895–902. [CrossRef]
33. Song, G.; Dai, Z.; Shen, Q.; Peng, X.; Zhang, M. Analysis of the changes in volatile compound and fatty acid profiles of fish oil in chemical refining process. *Eur. J. Lipid Sci. Technol.* **2017**, *120*, 1700219. [CrossRef]
34. FAO. *Fats and Fatty Acids in Human Nutrition: Report of an Expert Consultation*; FAO: Rome, Italy, 2010; p. 166.
35. Šimat, V.; Bogdanović, T.; Poljak, V.; Petričević, S. Changes in fatty acid composition, atherogenic and thrombogenic health lipid indices and lipid stability of bogue (*Boops boops* Linnaeus, 1758) during storage on ice: Effect of fish farming activities. *J. Food Comp. Anal.* **2015**, *40*, 120–125. [CrossRef]
36. Cameron-Smith, D.; Albert, B.B.; Cutfield, W.S. Fishing for answers: Is oxidation of fish oil supplements a problem? *J. Nutr. Sci.* **2015**, *4*, e36. [CrossRef]
37. Chung, H.; Choi, A.; Cho, I.H.; Kim, Y.-S. Changes in fatty acids and volatile components in mackerel by broiling. *Eur. J. Lipid Sci. Technol.* **2011**, *113*, 1481–1490. [CrossRef]
38. Jónsdóttir, R.; Bragadóttir, M.; Arnarson, G.Ö. Oxidatively derived volatile compounds in microencapsulated fish oil monitored by solid-phase microextraction (SPME). *J. Food Sci.* **2005**, *70*, 433–440. [CrossRef]
39. Roh, H.S.; Park, J.Y.; Park, S.Y.; Chun, B.S. Isolation of off-flavours from tuna fish oil using supercritical carbon dioxide. *Biotechnol. Bioprocess Eng.* **2006**, *11*, 496–502. [CrossRef]
40. Ghaly, A.E.; Dave, D.; Budge, S.; Brooks, M.S. Fish spoilage mechanisms and preservation techniques: Review. *Am. J Appl. Sci.* **2010**, *7*, 859–877. [CrossRef]
41. Giogios, I.; Grigorakis, K.; Nengas, I.; Papasolomontos, S.; Papaioannou, N.; Alexis, M.N. Fatty acid composition and volatile compounds of selected marine oils and meals. *J Sci. Food Agric.* **2008**, *89*, 88–100. [CrossRef]
42. De Oliveira, D.; Minozzo, M.G.; Licodiedoff, S.; Waszczynskyj, N. Physicochemical and sensory characterization of refined and deodorized tuna (*Thunnus albacares*) by-product oil obtained by enzymatic hydrolysis. *Food Chem.* **2016**, *207*, 187–194. [CrossRef]

MDPI
St. Alban-Anlage 66
4052 Basel
Switzerland
Tel. +41 61 683 77 34
Fax +41 61 302 89 18
www.mdpi.com

Foods Editorial Office
E-mail: foods@mdpi.com
www.mdpi.com/journal/foods

www.ingramcontent.com/pod-product-compliance
Lightning Source LLC
Chambersburg PA
CBHW062055220926
51850CB00010B/2778
9783039288519